STANDARD GRADE | CREDIT

# 2006

[BLANK PAGE]

# OFFICIAL SQA PAST PAPERS WITH ANSWERS

STANDARD GRADE | CREDIT

# CHEMISTRY
## 2006-2010

First exam published in 2006.
Published by Bright Red Publishing Ltd, 6 Stafford Street, Edinburgh EH3 7AU
tel: 0131 220 5804 fax: 0131 220 6710 info@brightredpublishing.co.uk www.brightredpublishing.co.uk

ISBN 978-1-84948-083-3

A CIP Catalogue record for this book is available from the British Library.

Bright Red Publishing is grateful to the copyright holders, as credited on the final page of the book, for permission to use their material.
Every effort has been made to trace the copyright holders and to obtain their permission for the use of copyright material.
Bright Red Publishing will be happy to receive information allowing us to rectify any error or omission in future editions.

FOR OFFICIAL USE

| | | | | | |
|---|---|---|---|---|---|
| | | | | | |

C

| | KU | PS |
|---|---|---|
| Total Marks | | |

# 0500/402

NATIONAL QUALIFICATIONS 2006

MONDAY, 8 MAY
10.50 AM – 12.20 PM

# CHEMISTRY
## STANDARD GRADE
### Credit Level

**Fill in these boxes and read what is printed below.**

Full name of centre

Town

Forename(s)

Surname

Date of birth
Day Month Year

Scottish candidate number

Number of seat

1 All questions should be attempted.

2 Necessary data will be found in the Data Booklet provided for Chemistry at Standard Grade and Intermediate 2.

3 The questions may be answered in any order but all answers are to be written in this answer book, and must be written clearly and legibly in ink.

4 Rough work, if any should be necessary, as well as the fair copy, is to be written in this book.

Rough work should be scored through when the fair copy has been written.

5 Additional space for answers and rough work will be found at the end of the book.

6 The size of the space provided for an answer should not be taken as an indication of how much to write. It is not necessary to use all the space.

7 Before leaving the examination room you must give this book to the invigilator. If you do not, you may lose all the marks for this paper.

SCOTTISH QUALIFICATIONS AUTHORITY

## PART 1

**In Questions 1 to 7 of this part of the paper, an answer is given by circling the appropriate letter (or letters) in the answer grid provided.**

**In some questions, two letters are required for full marks.**

**If more than the correct number of answers is given, marks will be deducted.**

**A total of 20 marks is available in this part of the paper.**

### SAMPLE QUESTION

| A | B | C |
|---|---|---|
| $CH_4$ | $H_2$ | $CO_2$ |
| D | E | F |
| CO | $C_2H_5OH$ | C |

(*a*) Identify the hydrocarbon.

| (A) | B | C |
|---|---|---|
| D | E | F |

The one correct answer to part (*a*) is A. This should be circled.

(*b*) Identify the **two** elements.

| A | (B) | C |
|---|---|---|
| D | E | (F) |

As indicated in this question, there are **two** correct answers to part (*b*). These are B and F. Both answers are circled.

If, after you have recorded your answer, you decide that you have made an error and wish to make a change, you should cancel the original answer and circle the answer you now consider to be correct. Thus, in part (*a*), if you want to change an answer A to an answer D, your answer sheet would look like this:

| ~~(A)~~ | B | C |
|---|---|---|
| (D) | E | F |

If you want to change back to an answer which has already been scored out, you should enter a tick (✓) in the box of the answer of your choice, thus:

| ✓ ~~(A)~~ | B | C |
|---|---|---|
| ~~(D)~~ | E | F |

DO NOT WRITE IN THIS MARGIN

Marks KU PS

**1.** The grid shows the names of some metals.

| A potassium | B platinum | C iron |
|---|---|---|
| D tin | E copper | F magnesium |

(*a*) Identify the metal used as a catalyst in the Ostwald Process.

| A | B | C |
|---|---|---|
| D | E | F |

**1**

(*b*) Identify the metal produced in a Blast Furnace.

| A | B | C |
|---|---|---|
| D | E | F |

**1**

(*c*) Identify the metal which has a density of 8·92 g/cm$^3$.

You may wish to use the data booklet to help you.

| A | B | C |
|---|---|---|
| D | E | F |

**1**

**(3)**

**[Turn over**

DO NOT WRITE IN THIS MARGIN

*Marks* KU PS

**2.** The structures of some hydrocarbons are shown in the grid below.

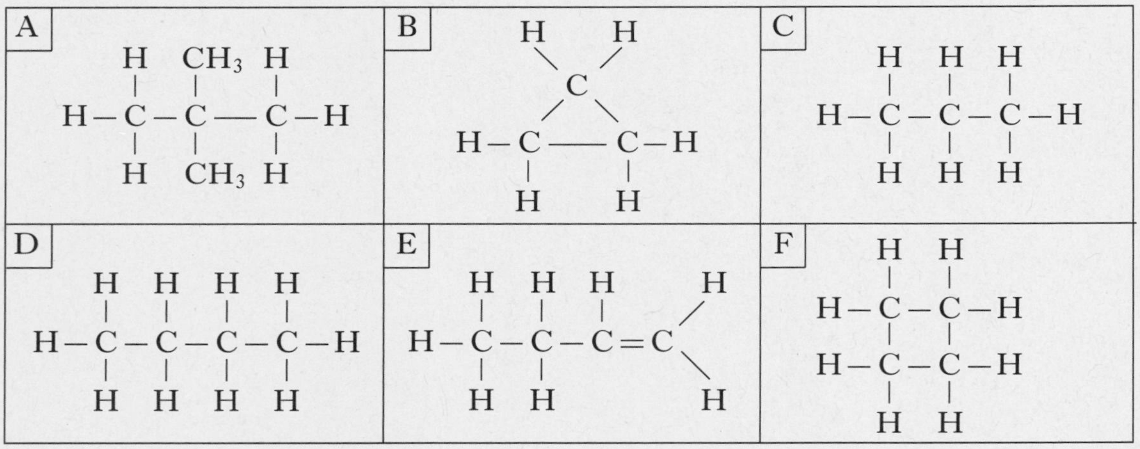


(*a*) Identify the **two** isomers.

| | | |
|---|---|---|
| A | B | C |
| D | E | F |

**1**

(*b*) Identify the hydrocarbon which is the first member of a homologous series.

| | | |
|---|---|---|
| A | B | C |
| D | E | F |

**1**

**(2)**

DO NOT WRITE IN THIS MARGIN

*Marks* KU PS

**3.** The grid shows the formulae for some gases.

| A | B | C |
|---|---|---|
| $O_2$ | $N_2$ | $CO$ |
| D | E | F |
| $SO_2$ | $NO_2$ | $CO_2$ |

(*a*) Identify the poisonous gas produced during the **incomplete combustion** of hydrocarbons.

| A | B | C |
|---|---|---|
| D | E | F |

**1**

(*b*) Identify the gas produced in air during a lightning storm.

| A | B | C |
|---|---|---|
| D | E | F |

**1**

(*c*) Identify the gas which is a reactant in the manufacture of ammonia (Haber Process).

| A | B | C |
|---|---|---|
| D | E | F |

**1**

**(3)**

**[Turn over**

Marks | KU | PS

**4.** The names of some oxides are shown in the grid.

| A | B | C |
|---|---|---|
| sodium oxide | potassium oxide | copper(II) oxide |
| D | E | F |
| carbon dioxide | zinc oxide | sulphur dioxide |

(*a*) Identify the **two** oxides which dissolve in water to form alkaline solutions.

| A | B | C |
|---|---|---|
| D | E | F |

**1**

(*b*) Identify the **two** oxides which are covalent.

| A | B | C |
|---|---|---|
| D | E | F |

**1**

**(2)**

Marks | KU | PS

**5.** The Periodic Table lists all known elements.

The grid shows the names of six common elements.

| A | B | C |
|---|---|---|
| oxygen | calcium | aluminium |
| D | E | F |
| sodium | magnesium | fluorine |

(*a*) Identify the **two** elements with similar chemical properties.

| A | B | C |
|---|---|---|
| D | E | F |

**1**

(*b*) Identify the element which can form ions with the same electron arrangement as argon.

| A | B | C |
|---|---|---|
| D | E | F |

**1**

(*c*) Identify the **two** elements which form an ionic compound with the formula of the type $\mathbf{XY_3}$, where **X** is the metal.

| A | B | C |
|---|---|---|
| D | E | F |

**1**

**(3)**

**[Turn over**

DO NOT WRITE IN THIS MARGIN

*Marks* KU PS

**6.** There are many different types of chemical reaction.

| A precipitation | B hydrolysis | C oxidation |
|---|---|---|
| D neutralisation | E condensation | F addition |

(*a*) Identify the type of chemical reaction that occurs when ethene reacts with hydrogen to form ethane.

| A | B | C |
|---|---|---|
| D | E | F |

**1**

(*b*) Identify the type of chemical reaction which occurs when a metal corrodes.

| A | B | C |
|---|---|---|
| D | E | F |

**1**

(*c*) Identify the **two** types of chemical reaction represented by the following equation.

$$Ba^{2+}(OH^-)_2(aq) + (H^+)_2SO_4^{2-}(aq) \longrightarrow Ba^{2+}SO_4^{2-}(s) + 2H_2O(\ell)$$

| A | B | C |
|---|---|---|
| D | E | F |

**2**

**(4)**

DO NOT WRITE IN THIS MARGIN

Marks KU PS

**7.** The grid shows pairs of chemicals.

| | |
|---|---|
| A<br>$CuO + C$ | B<br>$Na + H_2O$ |
| C<br>$Cu + NaNO_3$ | D<br>$C_5H_{12} + O_2$ |
| E<br>$Mg + H_2SO_4$ | F<br>$Ag + HCl$ |

(*a*) Which box contains a pair of chemicals that react to form water?

| | |
|---|---|
| A | B |
| C | D |
| E | F |

1

(*b*) Which **two** boxes contain pairs of chemicals that do **not** react together?

| | |
|---|---|
| A | B |
| C | D |
| E | F |

2

**(3)**

**[Turn over**

DO NOT WRITE IN THIS MARGIN

*Marks* KU PS

## PART 2

**A total of 40 marks is available in this part of the paper.**

**8.** Teflon is the brand name for the plastic, poly(tetrafluoroethene).

The structure of part of a poly(tetrafluoroethene) molecule is shown below.

```
   F   F   F   F   F   F
   |   |   |   |   |   |
 - C - C - C - C - C - C -
   |   |   |   |   |   |
   F   F   F   F   F   F
```

(*a*) Draw the full structural formula for the monomer used to make poly(tetrafluoroethene). **1**

(*b*) Teflon is a plastic which melts on heating.

What name is given to this type of plastic?

________________________________________ **1**

**(2)**

DO NOT WRITE IN THIS MARGIN

*Marks* KU PS

**9.** A sample of hydrogen was found to contain two different types of atom; $^{2}_{1}H$ and $^{1}_{1}H$.

(*a*) (i) What term is used to describe these different types of hydrogen atom?

______________________________ **1**

(ii) This sample of hydrogen has an average atomic mass of 1·1.

What is the mass number of the most common type of atom in this sample of hydrogen?

______________________________ **1**

(iii) Complete the table to show the number of protons and neutrons in each type of hydrogen atom.

| Type of atom | Number of protons | Number of neutrons |
|---|---|---|
| $^{2}_{1}H$ | | |
| $^{1}_{1}H$ | | |

**1**

(*b*) In a methane molecule ($CH_4$), hydrogen atoms form bonds with a carbon atom.

Draw a diagram to show the **shape** of a methane molecule.

**1**

**(4)**

**[Turn over**

DO NOT WRITE IN THIS MARGIN

*Marks* | KU | PS

**10.** A pupil set up the following experiment.

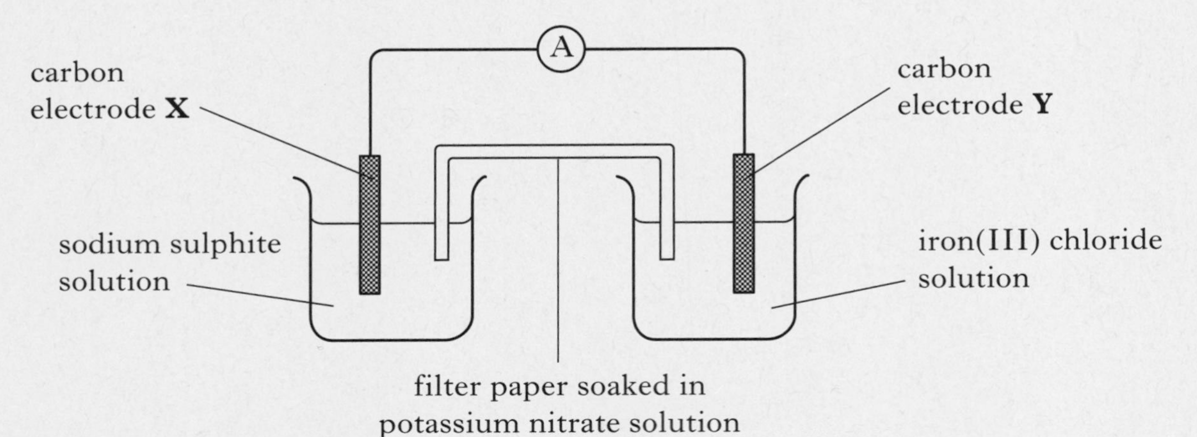


The reaction occurring at electrode **Y** is

$$Fe^{3+}(aq) + e^{-} \longrightarrow Fe^{2+}(aq)$$

(*a*) **On the diagram**, clearly mark the path and direction of electron flow. **1**

(*b*) Name the type of chemical reaction taking place at electrode **Y**.

________________________________ **1**

(*c*) After some time, ferroxyl indicator was added to the beaker containing electrode **Y**.

What colour would the ferroxyl indicator turn?

________________________________ **1**

**(3)**

DO NOT WRITE IN THIS MARGIN

*Marks* KU PS

**11.** Some types of steel can rust.

(*a*) Name the **two** substances which must be present for steel to rust.

________________________________________ **1**

(*b*) Paint containing "Red Lead" ($Pb_3O_4$) was used to protect steel from rusting.

Calculate the percentage, by mass, of lead in "Red Lead".

__________ % **2**

(*c*) Stainless steel is a type of steel which does not need protection. It contains chromium which forms a layer of chromium(III) oxide on the steel.

Write the formula for chromium(III) oxide.

________________________________________ **1**

**(4)**

**[Turn over**

DO NOT WRITE IN THIS MARGIN

*Marks* KU PS

**12.** Laura added the catalyst manganese dioxide to hydrogen peroxide solution and measured the volume of oxygen produced.

$$2H_2O_2(aq) \longrightarrow 2H_2O(\ell) + O_2(g)$$

Her results are shown in the table.

| **Time/s** | 0 | 10 | 30 | 40 | 50 | 60 |
|---|---|---|---|---|---|---|
| **Volume of oxygen/cm³** | 0 | 25 | 35 | 38 | 40 | 40 |

(*a*) Draw a line graph of the results.

*Use appropriate scales to fill most of the graph paper.*

(Additional graph paper, if required, will be found on page 25.)

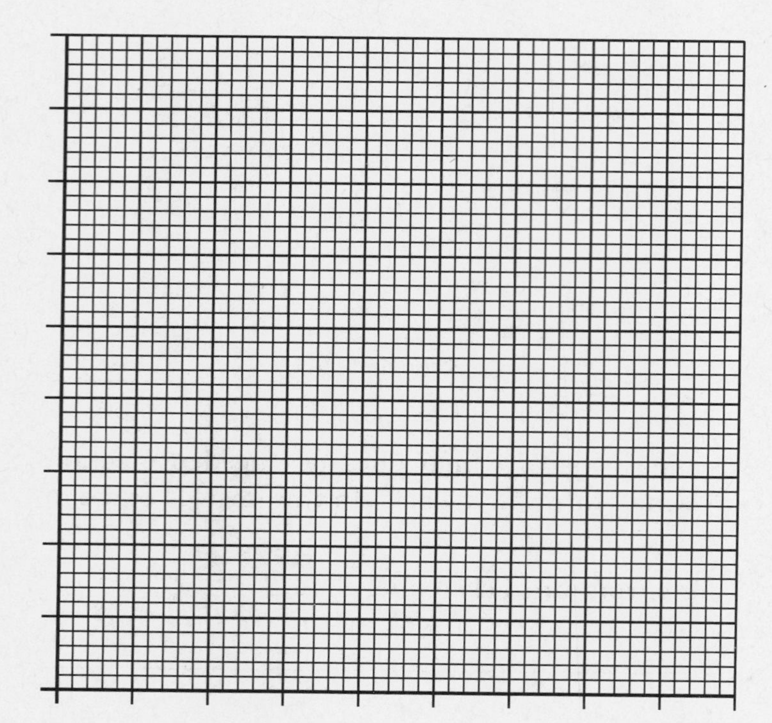

2

DO NOT WRITE IN THIS MARGIN

Marks | KU | PS

**12. (continued)**

(*b*) Using your graph, predict the volume of oxygen produced during the first 20 seconds.

_____________ $cm^3$ **1**

(*c*) Laura repeated the experiment at a higher temperature. She used the same volume and concentration of hydrogen peroxide solution.

Suggest a volume of oxygen produced during the first 30 seconds.

_____________ $cm^3$ **1**

**(4)**

**[Turn over**

DO NOT WRITE IN THIS MARGIN

Marks | KU | PS

**13.** Glucose is a carbohydrate.

(*a*) Name an isomer of glucose.

______________________________ **1**

(*b*) Glucose molecules join together to form starch in a polymerisation reaction.

Name the **type** of polymerisation reaction which takes place.

______________________________ **1**

(*c*) The diagram shows how glucose can be fermented to produce an alcohol. Carbon dioxide gas is also produced.

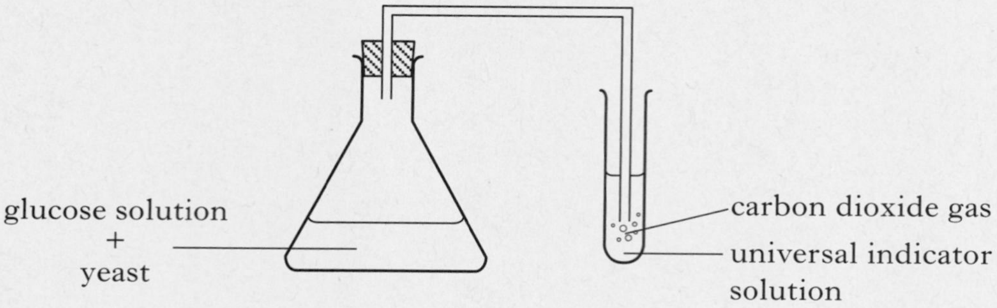

(i) What is the chemical name for the alcohol produced?

______________________________ **1**

(ii) Suggest the colour of the universal indicator solution after the carbon dioxide gas has been bubbled through it.

______________________________ **1**

**(4)**

DO NOT WRITE IN THIS MARGIN

*Marks* KU PS

**14.** Fuel cells produce electricity to power cars. The electricity is produced when hydrogen and oxygen react to form water.

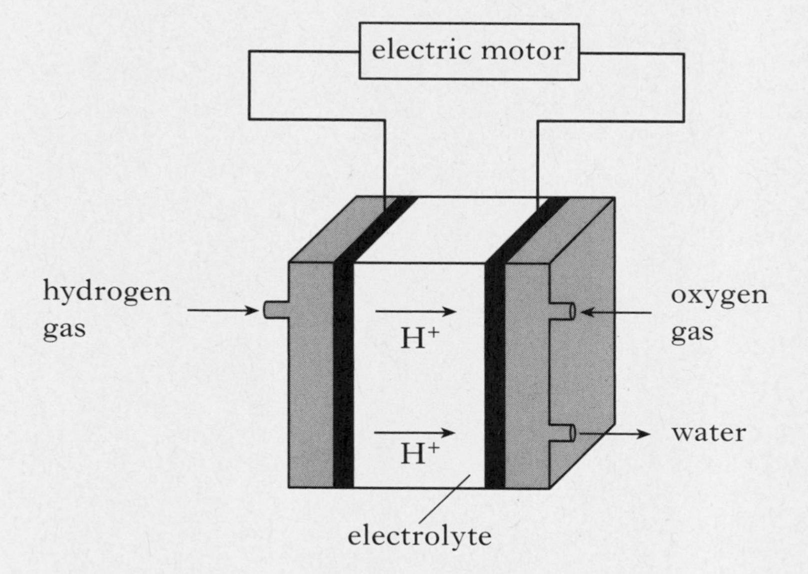


(*a*) Suggest a possible source of oxygen for use in the fuel cell.

______________________________ **1**

(*b*) Suggest an advantage in using fuel cells rather than petrol to power cars.

______________________________

______________________________ **1**

(*c*) Write the ion-electron equation for the formation of hydrogen ions.

You may wish to use the data booklet to help you.

______________________________ **1**

**(3)**

**[Turn over**

DO NOT WRITE IN THIS MARGIN

Marks | KU | PS

**15.** Dinitrogen monoxide can be used to increase power in racing cars.

Dinitrogen monoxide decomposes to form nitrogen and oxygen.

$$2N_2O(g) \longrightarrow 2N_2(g) + O_2(g)$$

dinitrogen monoxide

(*a*) Calculate the mass of oxygen produced, in grams, when 22 grams of dinitrogen monoxide decomposes.

Answer ______________ g **2**

(*b*) Tom set up the following experiment to compare the time taken for the burning candles in gas jars **A** and **B** to go out.

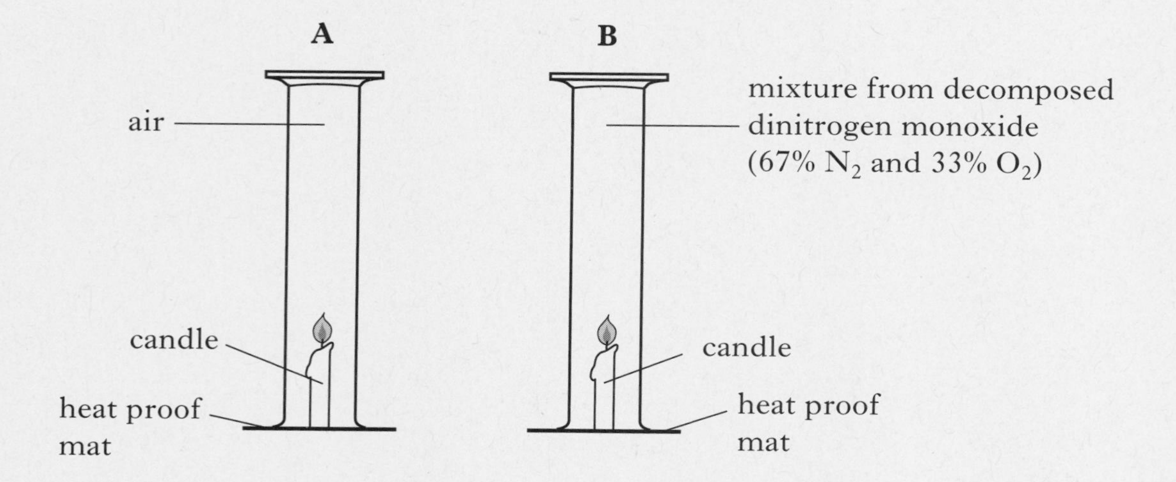


Circle the correct word in the table to show how the burning time of the candle in gas jar **B** compared to that in gas jar **A**.

| Candle | Burning time/s |
|---|---|
| **A** | 10 |
| **B** | same/longer/shorter |

**1**

**(3)**

DO NOT WRITE IN THIS MARGIN

*Marks* KU PS

**16.** Ammonia gas is produced when barium hydroxide reacts with ammonium chloride.

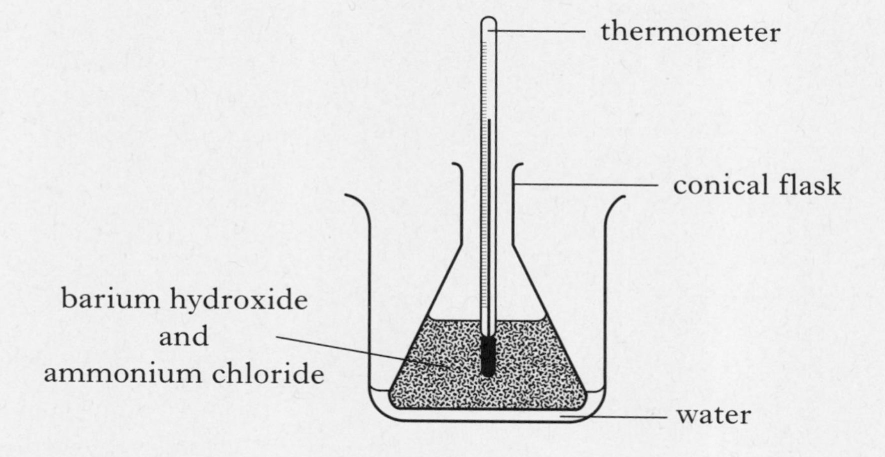


(*a*) The equation for the reaction which takes place is:

$$Ba(OH)_2 \quad + \quad NH_4Cl \quad \rightarrow \quad NH_3 \quad + \quad BaCl_2 \quad + \quad H_2O$$

Balance this equation. **1**

(*b*) Describe a test which would detect ammonia at the mouth of the flask.

__________________________________________________

__________________________________________________

__________________________________________________ **1**

(*c*) During the reaction the reading on the thermometer dropped from **25 °C** to **–5 °C**.

Suggest what would happen to the water in the beaker.

__________________________________________________

__________________________________________________

__________________________________________________ **1**

**(3)**

**[Turn over**

DO NOT WRITE IN THIS MARGIN

*Marks* KU PS

**17.** A class were given three chemicals labelled **X**, **Y** and **Z**.

The chemicals were glucose solution, copper chloride solution and dilute hydrochloric acid.

The apparatus below was used to help identify each solution.

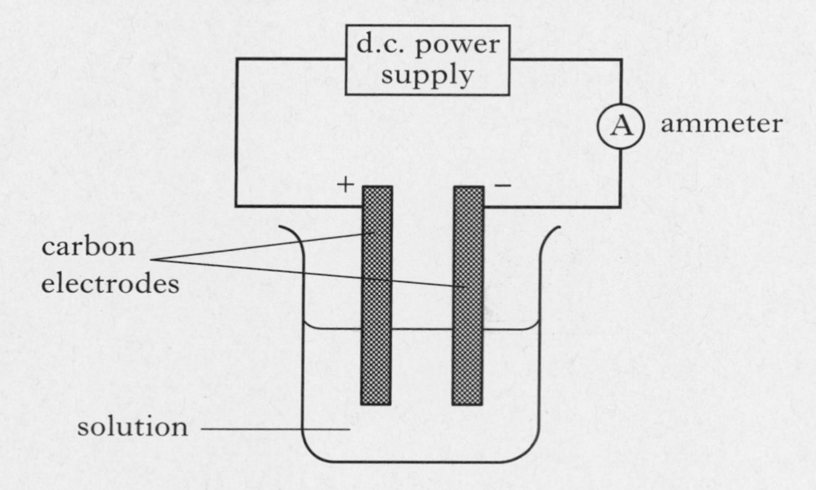


The class obtained the following results.

| Solution | Ammeter reading | Observations at electrodes |
|---|---|---|
| **X** | Yes | bubbles of gas formed at both electrodes |
| **Y** | Yes | brown solid formed at negative electrode |
| **Z** | No | no reaction |

(*a*) When electricity is passed through solutions **X** and **Y** they are broken up.

What term is used to describe this process?

____________________________________________ **1**

DO NOT WRITE IN THIS MARGIN

Marks | KU | PS

**17. (continued)**

(*b*) (i) Identify **X**.

________________________________ **1**

(ii) What type of bonding is present in **Z**?

________________________________ **1**

(iii) Describe what was **seen** at the positive electrode when electricity was passed through solution **Y**.

________________________________________________________ **1**

**(4)**

**[Turn over**

DO NOT WRITE IN THIS MARGIN

*Marks* KU PS

**18.** A pupil carried out a titration experiment to find the concentration of a potassium hydroxide solution.

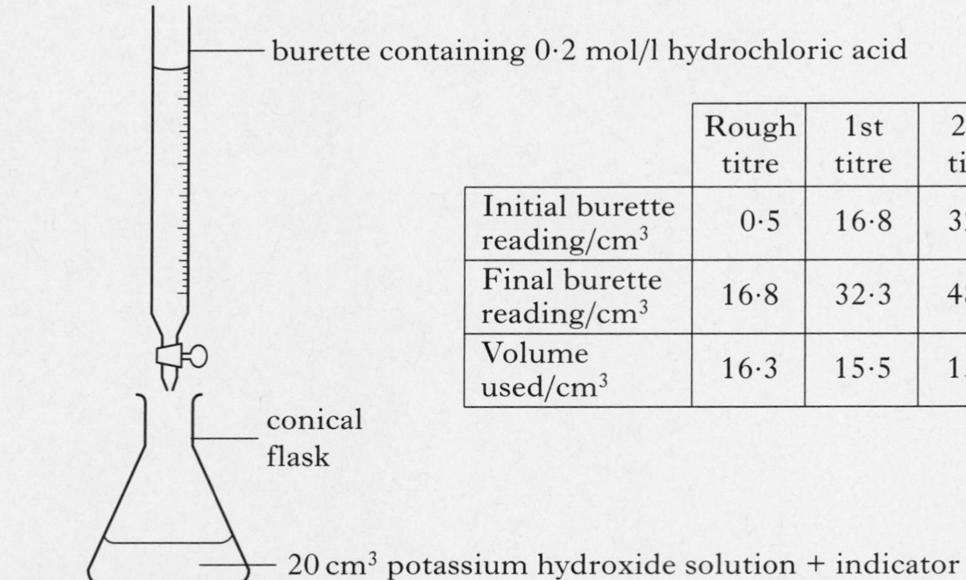


| | Rough titre | 1st titre | 2nd titre |
|---|---|---|---|
| Initial burette reading/cm³ | 0·5 | 16·8 | 32·3 |
| Final burette reading/cm³ | 16·8 | 32·3 | 48·0 |
| Volume used/cm³ | 16·3 | 15·5 | 15·7 |

The equation for the reaction is:

$$KOH(aq) + HCl(aq) \longrightarrow KCl(aq) + H_2O(\ell)$$

(*a*) Using the results in the table, calculate the **average** volume of hydrochloric acid required to neutralise the potassium hydroxide solution.

________________ cm³ **1**

DO NOT WRITE IN THIS MARGIN

*Marks* KU PS

**18. (continued)**

(*b*) Use the pupil's results to calculate the concentration, in mol/l, of the potassium hydroxide solution.

**Show your working clearly.**

________________ mol/l **2**

(*c*) The indicator was removed from the potassium chloride solution by filtering the solution through charcoal.

How would the pupil then obtain a sample of **solid** potassium chloride from the solution?

________________________________________

________________________________________

________________________________________

________________________________________ **1**

**(4)**

**[Turn over**

DO NOT WRITE IN THIS MARGIN

*Marks* KU PS

**19.** Esters are compounds used in perfumes.

Esters can be made when an alcohol reacts with a carboxylic acid.

```
     H               O  H                H        O  H           O
     |               ‖  |                |        ‖  |          / \
 H — C — O — H + H — O — C — C — H  →  H — C — O — C — C — H + H     H
     |                  |                |           |
     H                  H                H           H
 alcohol         carboxylic acid           ester            water
```

(*a*) Draw the **full** structural formula for the **ester** produced in the following reaction.

```
     H   H                 O   H   H
     |   |                 ‖   |   |
 H — C — C — O — H + H — O — C — C — C — H →
     |   |                     |   |
     H   H                     H   H
   alcohol              carboxylic acid
```

**1**

(*b*) After some time the esters in perfumes react with water and break down to form the alcohol and carboxylic acid again.

Suggest a name for the type of chemical reaction taking place.

______________________________________

**1**

**(2)**

[*END OF QUESTION PAPER*]

DO NOT WRITE IN THIS MARGIN

| KU | PS |
|---|---|

## ADDITIONAL SPACE FOR ANSWERS

ADDITIONAL GRAPH PAPER FOR QUESTION 12(*a*)

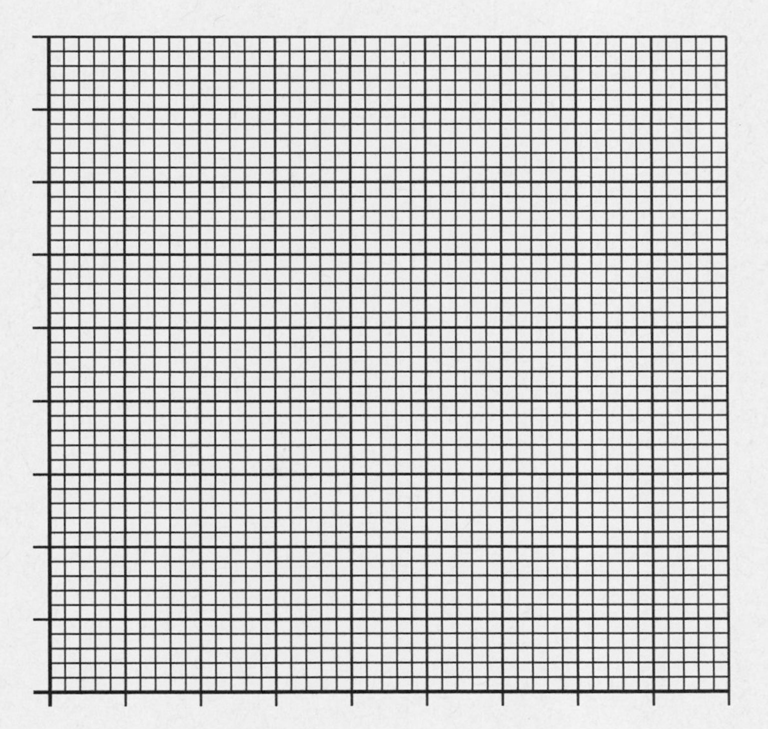

DO NOT WRITE IN THIS MARGIN

| KU | PS |
| --- | --- |

## ADDITIONAL SPACE FOR ANSWERS

STANDARD GRADE | CREDIT

# 2007

**[BLANK PAGE]**

FOR OFFICIAL USE

| | | | | | |
|---|---|---|---|---|---|
| | | | | | |

C

| | KU | PS |
|---|---|---|
| Total Marks | | |

# 0500/402

NATIONAL QUALIFICATIONS 2007

THURSDAY, 10 MAY
10.50 AM – 12.20 PM

# CHEMISTRY
## STANDARD GRADE
### Credit Level

**Fill in these boxes and read what is printed below.**

Full name of centre

Town

Forename(s)

Surname

Date of birth
Day Month Year

Scottish candidate number

Number of seat

1 All questions should be attempted.

2 Necessary data will be found in the Data Booklet provided for Chemistry at Standard Grade and Intermediate 2.

3 The questions may be answered in any order but all answers are to be written in this answer book, and must be written clearly and legibly in ink.

4 Rough work, if any should be necessary, as well as the fair copy, is to be written in this book.

Rough work should be scored through when the fair copy has been written.

5 Additional space for answers and rough work will be found at the end of the book.

6 The size of the space provided for an answer should not be taken as an indication of how much to write. It is not necessary to use all the space.

7 Before leaving the examination room you must give this book to the invigilator. If you do not, you may lose all the marks for this paper.

SA 0500/402 6/26770

## PART 1

**In Questions 1 to 9 of this part of the paper, an answer is given by circling the appropriate letter (or letters) in the answer grid provided.**

**In some questions, two letters are required for full marks.**

**If more than the correct number of answers is given, marks will be deducted.**

**A total of 20 marks is available in this part of the paper.**

### SAMPLE QUESTION

| A $CH_4$ | B $H_2$ | C $CO_2$ |
|---|---|---|
| D $CO$ | E $C_2H_5OH$ | F $C$ |

(*a*) Identify the hydrocarbon.

| (A) | B | C |
|---|---|---|
| D | E | F |

The one correct answer to part (*a*) is A. This should be circled.

(*b*) Identify the **two** elements.

| A | (B) | C |
|---|---|---|
| D | E | (F) |

As indicated in this question, there are **two** correct answers to part (*b*). These are B and F. Both answers are circled.

If, after you have recorded your answer, you decide that you have made an error and wish to make a change, you should cancel the original answer and circle the answer you now consider to be correct. Thus, in part (*a*), if you want to change an answer A to an answer D, your answer sheet would look like this:

| (A) crossed out | B | C |
|---|---|---|
| (D) | E | F |

If you want to change back to an answer which has already been scored out, you should enter a tick (✓) in the box of the answer of your choice, thus:

| ✓ (A) crossed out | B | C |
|---|---|---|
| (D) crossed out | E | F |

DO NOT WRITE IN THIS MARGIN

*Marks* KU PS

**1.**

**Testing gases**

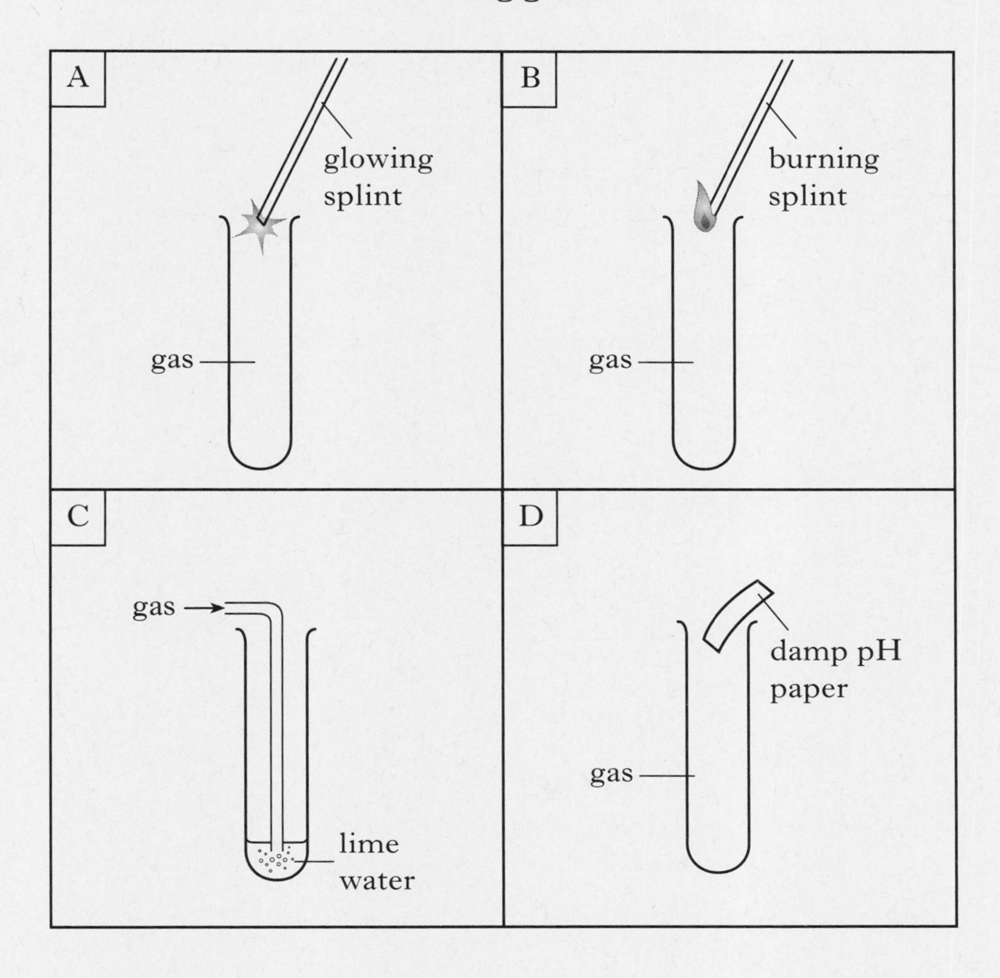


(*a*) Identify the test for oxygen gas.

| A | B |
|---|---|
| C | D |

**1**

(*b*) Identify a test for ammonia gas.

| A | B |
|---|---|
| C | D |

**1**

**(2)**

**[Turn over**

DO NOT WRITE IN THIS MARGIN

*Marks* KU PS

**2.** Zinc and magnesium both react with dilute hydrochloric acid.

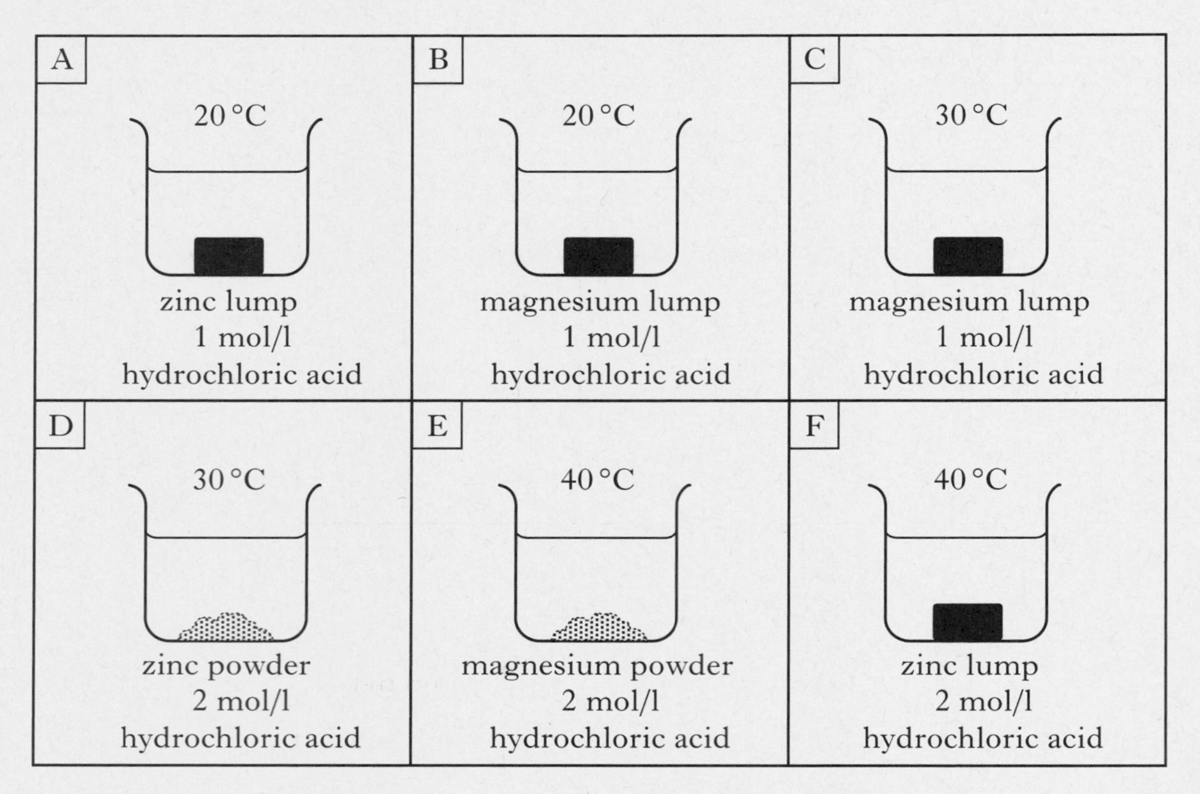


(*a*) Identify the experiment with the **slowest** rate of reaction.

| A | B | C |
|---|---|---|
| D | E | F |

**1**

(*b*) Identify the **two** experiments which could be used to investigate the effect of temperature on the rate of reaction.

| A | B | C |
|---|---|---|
| D | E | F |

**1**

**(2)**

DO NOT WRITE IN THIS MARGIN

Marks KU PS

**3.** Distillation of crude oil produces several fractions.

crude oil

| *Fraction* | *Number of carbon atoms per molecule* |
|---|---|
| A | 1–4 |
| B | 4–10 |
| C | 10–16 |
| D | 16–20 |
| E | 20+ |

(*a*) Identify the fraction which is used to tar roads.

| A |
|---|
| B |
| C |
| D |
| E |

**1**

(*b*) Identify the fraction with the lowest boiling point.

| A |
|---|
| B |
| C |
| D |
| E |

**1**

**(2)**

**[Turn over**

DO NOT WRITE IN THIS MARGIN

Marks KU PS

**4.** The structural formulae for some hydrocarbons are shown below.

| A | B | C |
|---|---|---|
| $CH_3$ H<br>C = C<br>H H | H H<br>C<br>H C — C H<br>H H | H $CH_3$<br>H – C — C – H<br>$CH_3$ H |
| **D** | **E** | **F** |
| H H<br>H – C – C – H<br>H – C – C – H<br>H H | $CH_3$ H<br>H – C — C – H<br>H H | $CH_3$ H<br>C = C<br>H $CH_3$ |

(*a*) Identify the hydrocarbon which could be used to make poly(butene).

| A | B | C |
|---|---|---|
| D | E | F |

1

(*b*) Identify the **two** hydrocarbons with the general formula $C_nH_{2n}$ which do **not** react quickly with hydrogen.

| A | B | C |
|---|---|---|
| D | E | F |

1

(2)

DO NOT WRITE IN THIS MARGIN

Marks KU PS

**5.** The table contains information about some substances.

| *Substance* | *Melting point/ °C* | *Boiling point/ °C* | *Conducts as a solid* | *Conducts as a liquid* |
|---|---|---|---|---|
| A | 1700 | 2230 | no | no |
| B | 605 | 1305 | no | yes |
| C | –13 | 77 | no | no |
| D | 801 | 1413 | no | yes |
| E | 181 | 1347 | yes | yes |
| F | –39 | 357 | yes | yes |

(*a*) Identify the substance which exists as covalent molecules.

| A |
|---|
| B |
| C |
| D |
| E |
| F |

**1**

(*b*) Identify the metal which is liquid at 25 °C.

| A |
|---|
| B |
| C |
| D |
| E |
| F |

**1**

**(2)**

**[Turn over**

DO NOT WRITE IN THIS MARGIN

Marks | KU | PS

**6.** Equations are used to represent chemical reactions.

| | |
|---|---|
| A | $Sn(s) \longrightarrow Sn^{2+}(aq) + 2e^-$ |
| B | $Cu^{2+}(aq) + 2e^- \longrightarrow Cu(s)$ |
| C | $H^+(aq) + OH^-(aq) \longrightarrow H_2O(\ell)$ |
| D | $2Mg(s) + O_2(g) \longrightarrow 2MgO(s)$ |
| E | $SO_2(g) + H_2O(\ell) \longrightarrow 2H^+(aq) + SO_3^{2-}(aq)$ |

(*a*) Identify the equation which represents the formation of acid rain.

| A |
|---|
| B |
| C |
| D |
| E |

1

(*b*) Identify the equation which represents neutralisation.

| A |
|---|
| B |
| C |
| D |
| E |

1

(*c*) Identify the **two** equations in which a substance is oxidised.

| A |
|---|
| B |
| C |
| D |
| E |

2

**(4)**

DO NOT WRITE IN THIS MARGIN

*Marks* KU PS

**7.** A student made some statements about the particles found in atoms.

| | |
|---|---|
| A | It has a positive charge. |
| B | It has a negative charge. |
| C | It has a relative mass of almost zero. |
| D | It has a relative mass of 1. |
| E | It is found inside the nucleus. |
| F | It is found outside the nucleus. |

Identify the **two** statements which apply to **both** a proton and a neutron.

| |
|---|
| A |
| B |
| C |
| D |
| E |
| F |

**(2)**

**8.** A student made some statements about the reaction of silver(I) oxide with excess dilute hydrochloric acid.

| | |
|---|---|
| A | The concentration of hydrogen ions increases. |
| B | Carbon dioxide gas is produced. |
| C | An insoluble salt is produced. |
| D | Hydrogen gas is produced. |
| E | Water is produced. |

Identify the **two** correct statements.

| |
|---|
| A |
| B |
| C |
| D |
| E |

**(2)**

DO NOT WRITE IN THIS MARGIN

*Marks* KU PS

**9.** When two different electrodes are joined in a cell, a chemical reaction takes place and a voltage is produced.

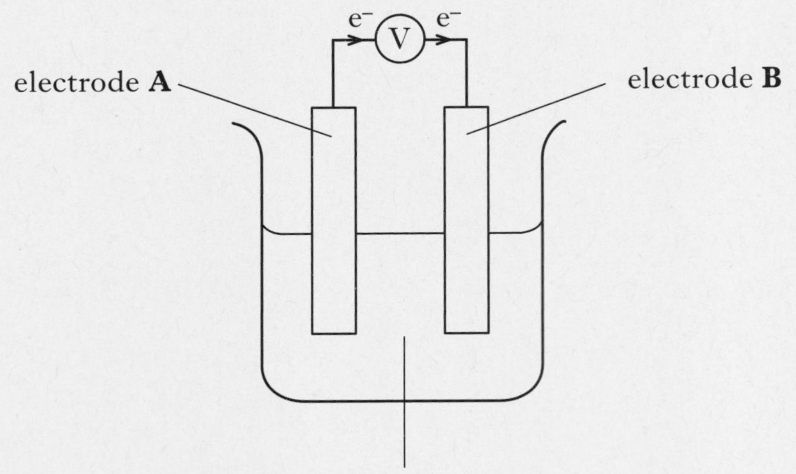


sodium chloride solution and ferroxyl indicator

| | ***Electrode A*** | ***Electrode B*** |
|---|---|---|
| **A** | magnesium | iron |
| **B** | iron | carbon |
| **C** | iron | aluminium |
| **D** | iron | copper |
| **E** | lead | iron |

Which **two** pairs of electrodes will produce a flow of electrons in the same direction as shown in the diagram and would produce a blue colour around electrode **A**?

You may wish to use the data booklet to help you.

| A |
|---|
| B |
| C |
| D |
| E |

**(2)**

**[Turn over for Part 2 on *Page twelve***

DO NOT WRITE IN THIS MARGIN

*Marks* KU PS

## PART 2

**A total of 40 marks is available in this part of the paper.**

**10.** A sample of silver was found to contain two isotopes, $^{107}_{47}Ag$ and $^{109}_{47}Ag$.

(*a*) This sample of silver has an average atomic mass of 108.

What does this indicate about the amount of each isotope in this sample?

________________________________________

________________________________________ **1**

(*b*) Complete the table to show the number of each type of particle in a $^{107}_{47}Ag^{+}$ ion.

| ***Particle*** | ***Number*** |
|---|---|
| proton | |
| neutron | |
| electron | |

**2**

(*c*) Silver can be displaced from a solution of silver(I) nitrate.

$$AgNO_3(aq) \quad + \quad Cu(s) \quad \longrightarrow \quad Ag(s) \quad + \quad Cu(NO_3)_2(aq)$$

(i) Balance this equation. **1**

(ii) Name a metal which would **not** displace silver from silver(I) nitrate.

You may wish to use the data booklet to help you.

________________________________________ **1**

**(5)**

DO NOT WRITE IN THIS MARGIN

*Marks* KU PS

**11.** Alkanoic acids are a family of compounds which contain the $-C(=O)-O-H$ group.

The **full** structural formulae for the first three members are shown.

$$H-C(=O)-O-H$$

methanoic acid

$$H-\underset{H}{\overset{H}{C}}-C(=O)-O-H$$

ethanoic acid

$$H-\underset{H}{\overset{H}{C}}-\underset{H}{\overset{H}{C}}-C(=O)-O-H$$

propanoic acid

(*a*) Draw the **full** structural formula for the alkanoic acid containing 4 carbon atoms. **1**

(*b*) The table gives information on some alkanoic acids.

| *Acid* | *Boiling point/ °C* |
|---|---|
| methanoic acid | 101 |
| ethanoic acid | 118 |
| propanoic acid | 141 |
| butanoic acid | 164 |

(i) Using this information, make a general statement linking the boiling point to the number of carbon atoms.

______________________________________________

______________________________________________

______________________________________________ **1**

(ii) Predict the boiling point of pentanoic acid.

__________ °C **1**

**(3)**

[Turn over

DO NOT WRITE IN THIS MARGIN

Marks KU PS

**12.** Ammonia is made when nitrogen and hydrogen react together.

The table below shows the percentage yields obtained when nitrogen and hydrogen react at different pressures.

| ***Pressure/atmospheres*** | ***Percentage yield of ammonia*** |
|---|---|
| 25 | 28 |
| 50 | 40 |
| 100 | 53 |
| 200 | 67 |
| 400 | 80 |

(*a*) Draw a line graph of percentage yield against pressure.

*Use appropriate scales to fill most of the graph paper.*

(Additional graph paper, if required, will be found on page 27.)

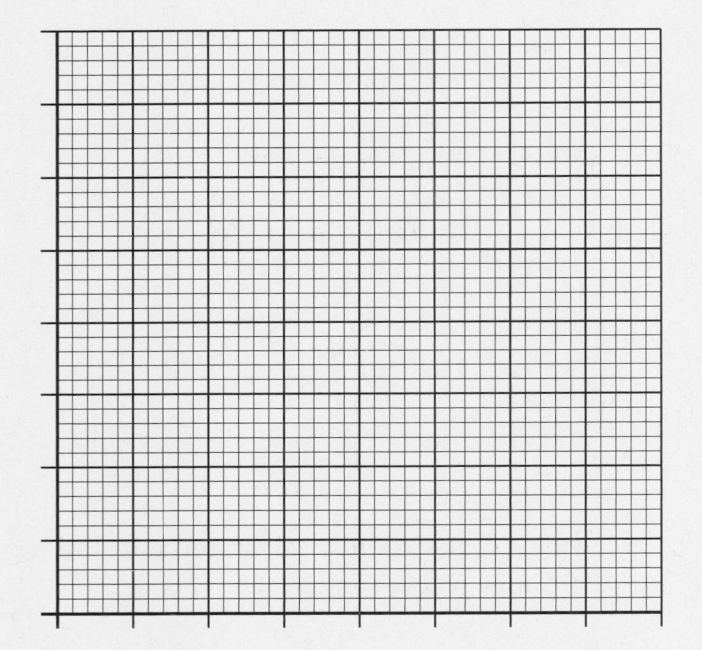

2

(*b*) Use your graph to estimate the percentage yield of ammonia at 150 atmospheres.

________________________________________ 1

DO NOT WRITE IN THIS MARGIN

*Marks* KU PS

**12. (continued)**

(*c*) Ammonia can be produced in the lab by heating an ammonium compound with soda lime.

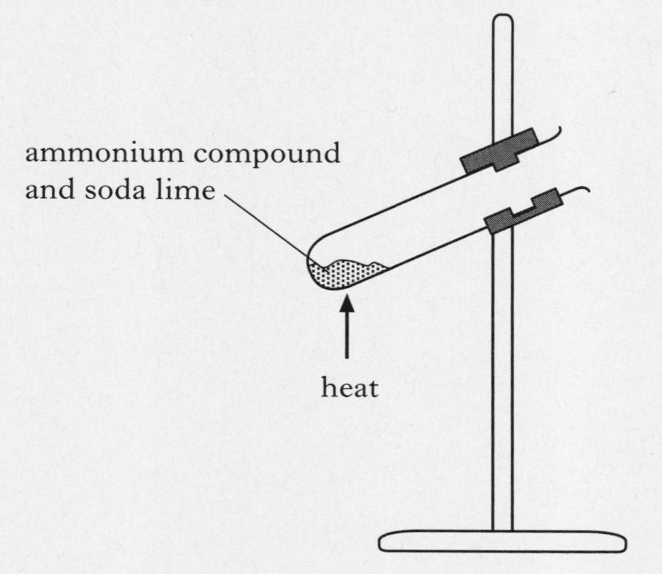


In order to produce ammonia, what **type** of compound must soda lime be?

________________________________________ **1**

**(4)**

**[Turn over**

DO NOT WRITE IN THIS MARGIN

*Marks* KU PS

**13.** Starch and sucrose can be hydrolysed to produce simple sugars.

Chromatography is a technique which can be used to identify the sugars produced.

Samples of known sugar solutions are spotted on the base line. The solvent travels up the paper carrying spots of sugars at different rates.

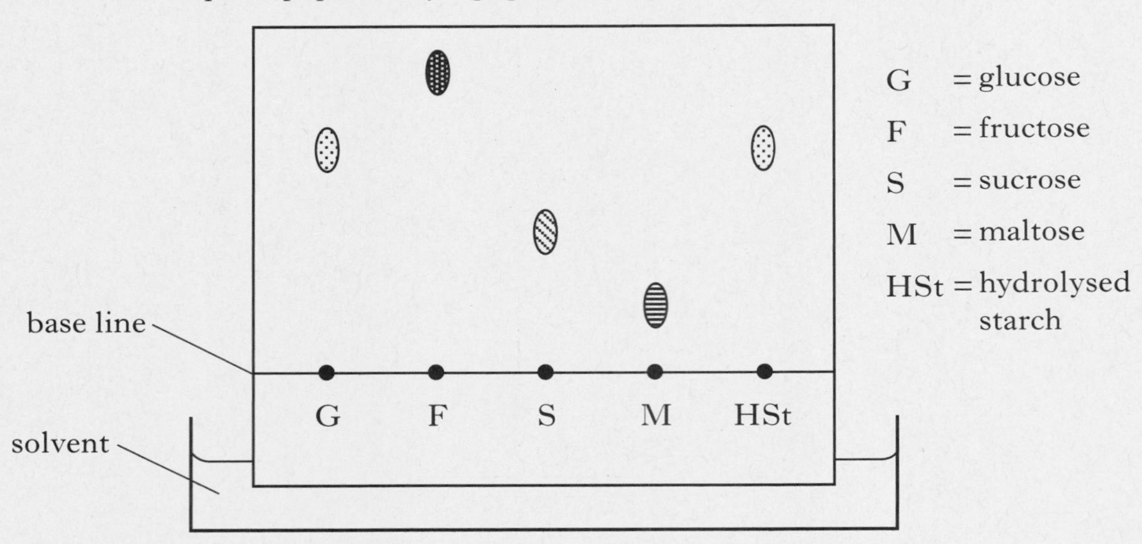


The diagram above shows that **only glucose** is produced when starch is hydrolysed.

(*a*) The chromatogram below can be used to identify the simple sugars produced when sucrose is hydrolysed.

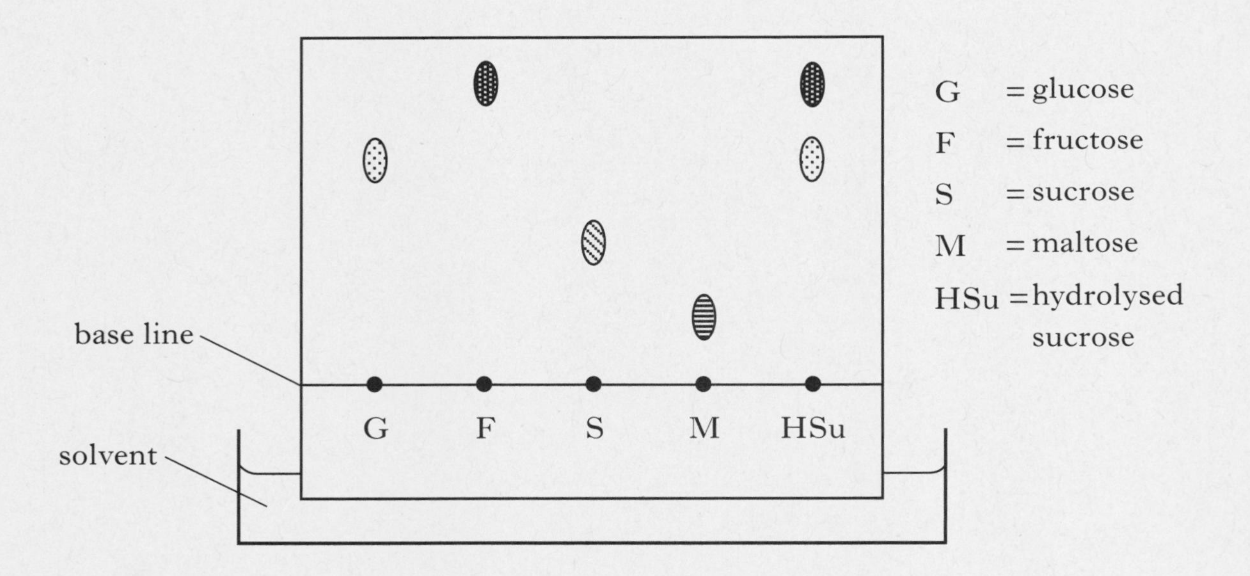


Name the sugars produced when sucrose is hydrolysed.

______________________________________________ **1**

DO NOT WRITE IN THIS MARGIN

*Marks* KU PS

**13. (continued)**

(*b*) What **type** of substance, present in the digestive system, acts as a catalyst in the hydrolysis of sucrose?

____________________________________________ **1**

**(2)**

**[Turn over**

DO NOT WRITE IN THIS MARGIN

*Marks* | KU | PS

**14.** Cars made from steel can be protected from rusting in a number of ways.

(*a*) Circle the correct word to complete the sentence below.

Steel does not rust when attached to the { negative / positive } terminal of a car battery. **1**

(*b*) The steel body of the car can be coated by dipping it in molten zinc.

(i) What name is given to this process?

________________________________________ **1**

(ii) Explain why the steel does **not** rust even when the zinc coating is scratched.

________________________________________

________________________________________

________________________________________ **1**

**(3)**

DO NOT WRITE IN THIS MARGIN

*Marks* KU PS

**15.** The atoms in a chlorine molecule are held together by a covalent bond. A covalent bond is a shared pair of electrons.

The chlorine molecule can be represented as

● = electron

(*a*) Showing **all** outer electrons, draw a similar diagram to represent a molecule of hydrogen chloride, HCl. **1**

(*b*) In forming covalent bonds, why do atoms share electrons?

______________________________________________

______________________________________________

______________________________________________ **1**

**(2)**

**[Turn over**

*Marks* KU PS

**16.** Ethanol is the alcohol found in alcoholic drinks.

It can be produced as shown in the diagram.

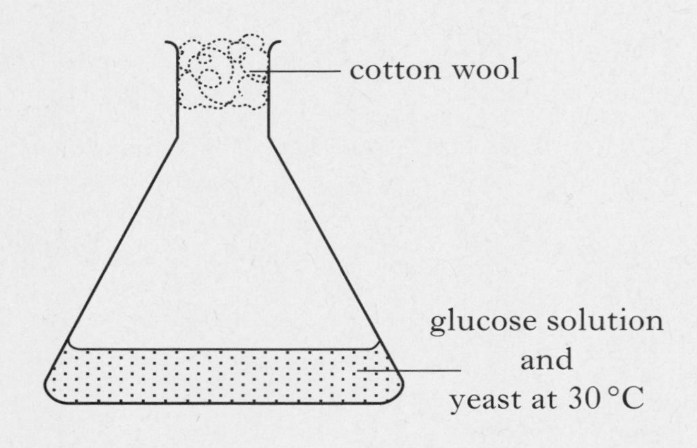


(*a*) (i) Name the type of chemical reaction taking place in the flask.

________________________________________ **1**

(ii) What would happen to the rate of the reaction if the experiment above was repeated at 50 °C?

________________________________________ **1**

(*b*) In industry, alcohols can be produced from alkenes as shown in the example below.

```
                                              H   H   H
                                              |   |   |
                                          H — C — C — C — H
                                              |   |   |
    H   H   H                                 H   H   O-H
    |   |   |              O          ↗       propan-1-ol
H — C — C = C — H  +  H       H
    |                                 ↘       H   H   H
    H                                         |   |   |
  propene                water            H — C — C — C — H
                                              |   |   |
                                              H  O-H  H
                                              propan-2-ol
```

(i) Name the type of chemical reaction taking place.

________________________________________ **1**

DO NOT WRITE IN THIS MARGIN

*Marks* KU PS

**16. (*b*) (continued)**

(ii) What **term** is used to describe a pair of alcohols like propan-1-ol and propan-2-ol?

______________________________ **1**

(iii) Propan-1-ol and propan-2-ol have different boiling points.

Name the process which could be used to separate a mixture of these alcohols.

______________________________ **1**

**(5)**

**[Turn over**

DO NOT WRITE IN THIS MARGIN

Marks KU PS

**17.** The table contains information on minerals.

| ***Mineral*** | ***Formula*** |
|---|---|
| cinnabar | $HgS$ |
| fluorite | $CaF_2$ |
| gibbsite | $Al(OH)_3$ |
| haematite | $Fe_2O_3$ |
| zinc blende | $ZnS$ |

(*a*) State the chemical name for zinc blende.

________________________________________ **1**

(*b*) Name the salt formed when gibbsite reacts with dilute hydrochloric acid.

________________________________________ **1**

(*c*) Calculate the percentage, by mass, of calcium in fluorite ($CaF_2$).

**Show your working clearly**.

________ % **2**

(*d*) Iron metal can be extracted from haematite ($Fe_2O_3$) by heating with carbon monoxide. Carbon dioxide is also produced.

Write an equation, using **symbols** and **formulae**, for this reaction.

There is no need to balance it.

________________________________________ **1**

(*e*) Name a metal which can be extracted from its ore by heat alone.

________________________________________ **1**

**(6)**

DO NOT WRITE IN THIS MARGIN

*Marks* KU PS

**18.** Nylon is a polymer with many uses.

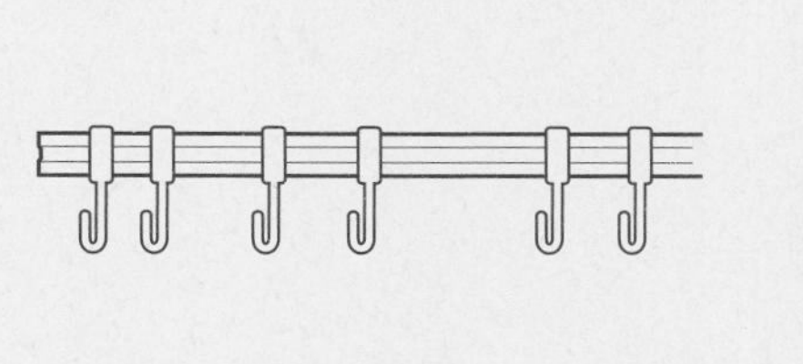

curtain rail

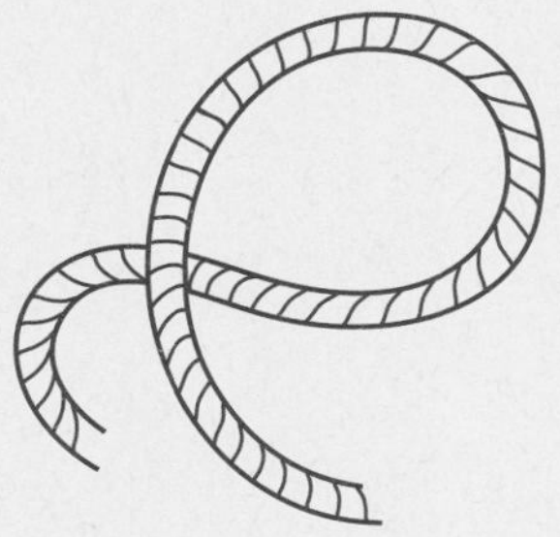

rope

jacket

(*a*) Nylon is a thermoplastic polymer.

What does thermoplastic mean?

____________________________________________

____________________________________________ **1**

(*b*) Nylon is a polymer made from two different monomers as shown.

$$H-N(H)-(CH_2)_6-N(H)-H \quad H-O-C(=O)-(CH_2)_4-C(=O)-O-H$$

$$\downarrow$$

$$-N(H)-(CH_2)_6-N(H)-C(=O)-(CH_2)_4-C(=O)- \quad + \quad H-O-H$$

nylon

During the polymerisation reaction, water is also produced.

Suggest a name for this **type** of polymerisation.

____________________________________________ **1**

**(2)**

**[Turn over**

DO NOT WRITE IN THIS MARGIN

*Marks* KU PS

**19.** Many ionic compounds are coloured.

| *Compound* | *Colour* |
|---|---|
| nickel(II) nitrate | green |
| nickel(II) sulphate | green |
| potassium permanganate | purple |
| potassium sulphate | colourless |

(*a*) Using the information in the table, state the colour of the potassium ion.

______________________________________________ **1**

(*b*) Write the **ionic** formula for nickel(II) nitrate.

______________________________________________ **1**

(*c*) A student set up the following experiment to investigate the colour of the ions in copper(II) chromate.

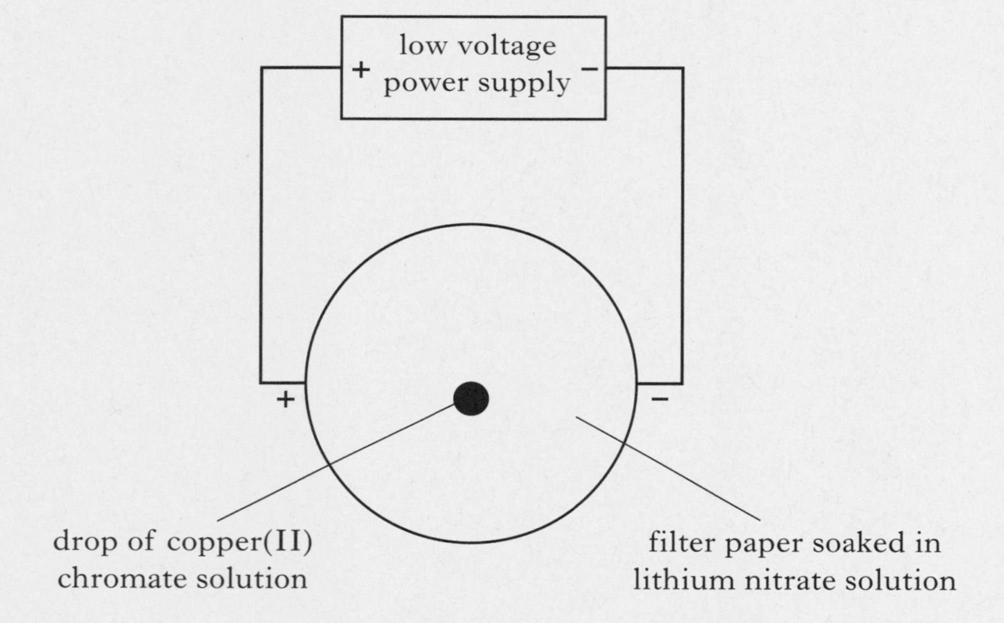


The student made the following observation.

| *Observation* |
|---|
| yellow colour moves to the positive electrode |
| blue colour moves to the negative electrode |

DO NOT WRITE IN THIS MARGIN

*Marks* KU PS

**19. (c) (continued)**

(i) State the colour of the chromate ion.

______________________________ **1**

(ii) Lithium nitrate solution is used as the electrolyte.

What is the purpose of an electrolyte?

______________________________

______________________________

______________________________ **1**

(iii) Suggest why lithium phosphate can **not** be used as the electrolyte in this experiment.

You may wish to use the data booklet to help you.

______________________________

______________________________

______________________________ **1**

**(5)**

**[Turn over**

DO NOT WRITE IN THIS MARGIN

Marks | KU | PS

**20.** Indigestion is caused by excess acid in the stomach. Indigestion remedies containing calcium carbonate neutralise some of this acid.

Christine carried out an experiment to find the mass of calcium carbonate required to neutralise a dilute hydrochloric acid solution.

She added calcium carbonate until all the acid had been used up.

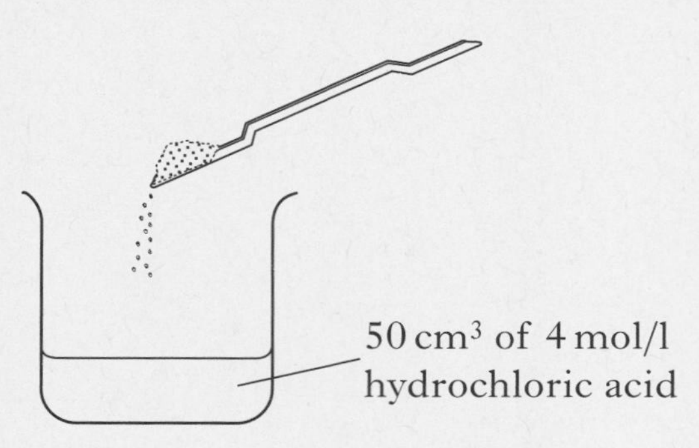


(*a*) Calculate the number of moles of dilute hydrochloric acid used in the experiment.

________________ mol **1**

(*b*) The equation for the reaction is

$$CaCO_3(s) + 2HCl(aq) \longrightarrow CaCl_2(aq) + H_2O(\ell) + CO_2(g)$$

(i) Using your answer from part (*a*), calculate the number of moles of calcium carbonate required to neutralise the dilute hydrochloric acid.

________________ mol **1**

(ii) Using your answer from part (*b*)(i), calculate the **mass** of calcium carbonate ($CaCO_3$) required to neutralise the acid.

________________ g **1**

**(3)**

[*END OF QUESTION PAPER*]

DO NOT WRITE IN THIS MARGIN

| KU | PS |
|---|---|

## ADDITIONAL SPACE FOR ANSWERS

ADDITIONAL GRAPH PAPER FOR QUESTION 12(*a*)

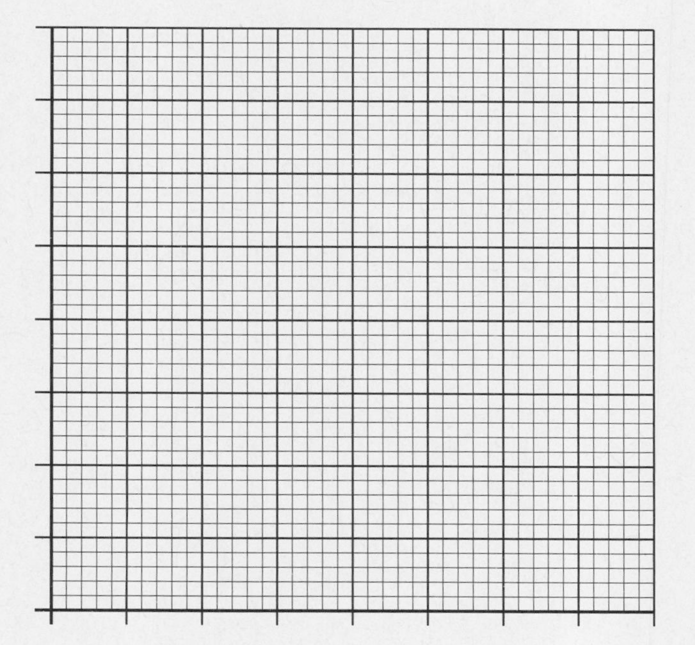

[BLANK PAGE]

STANDARD GRADE | CREDIT

# 2008

[BLANK PAGE]

FOR OFFICIAL USE

| | | | | | |
|---|---|---|---|---|---|
| | | | | | |

C

| | KU | PS |
|---|---|---|
| Total Marks | | |

# 0500/402

NATIONAL QUALIFICATIONS 2008

THURSDAY, 1 MAY
10.50 AM – 12.20 PM

# CHEMISTRY
## STANDARD GRADE
### Credit Level

**Fill in these boxes and read what is printed below.**

Full name of centre

Town

Forename(s)

Surname

Date of birth
Day Month Year

Scottish candidate number

Number of seat

1 All questions should be attempted.

2 Necessary data will be found in the Data Booklet provided for Chemistry at Standard Grade and Intermediate 2.

3 The questions may be answered in any order but all answers are to be written in this answer book, and must be written clearly and legibly in ink.

4 Rough work, if any should be necessary, as well as the fair copy, is to be written in this book.
Rough work should be scored through when the fair copy has been written.

5 Additional space for answers and rough work will be found at the end of the book.

6 The size of the space provided for an answer should not be taken as an indication of how much to write. It is not necessary to use all the space.

7 Before leaving the examination room you must give this book to the invigilator. If you do not, you may lose all the marks for this paper.

SA 0500/402 6/26070

## PART 1

**In Questions 1 to 9 of this part of the paper, an answer is given by circling the appropriate letter (or letters) in the answer grid provided.**

**In some questions, two letters are required for full marks.**

**If more than the correct number of answers is given, marks will be deducted.**

**A total of 20 marks is available in this part of the paper.**

### SAMPLE QUESTION

| A | B | C |
|---|---|---|
| $CH_4$ | $H_2$ | $CO_2$ |
| D | E | F |
| CO | $C_2H_5OH$ | C |

(*a*) Identify the hydrocarbon.

| (A) | B | C |
|---|---|---|
| D | E | F |

The one correct answer to part (*a*) is A. This should be circled.

(*b*) Identify the **two** elements.

| A | (B) | C |
|---|---|---|
| D | E | (F) |

As indicated in this question, there are **two** correct answers to part (*b*). These are B and F. Both answers are circled.

If, after you have recorded your answer, you decide that you have made an error and wish to make a change, you should cancel the original answer and circle the answer you now consider to be correct. Thus, in part (*a*), if you want to change an answer A to an answer D, your answer sheet would look like this:

| ~~(A)~~ | B | C |
|---|---|---|
| (D) | E | F |

If you want to change back to an answer which has already been scored out, you should enter a tick (✓) in the box of the answer of your choice, thus:

| ✓ ~~(A)~~ | B | C |
|---|---|---|
| ~~(D)~~ | E | F |

DO NOT WRITE IN THIS MARGIN

*Marks* KU PS

**1.** The formulae of some gases are shown in the grid.

| A | B | C |
|---|---|---|
| $H_2$ | $N_2$ | CO |
| D | E | F |
| $O_2$ | $CO_2$ | $NO_2$ |

(*a*) Identify the toxic gas produced during the burning of plastics.

| A | B | C |
|---|---|---|
| D | E | F |

**1**

(*b*) Identify the gas which makes up approximately 80% of air.

| A | B | C |
|---|---|---|
| D | E | F |

**1**

(*c*) Identify the gas used up during respiration.

| A | B | C |
|---|---|---|
| D | E | F |

**1**

**(3)**

**[Turn over**

DO NOT WRITE IN THIS MARGIN

*Marks* KU PS

**2.** A student carried out several experiments with metals and acids.

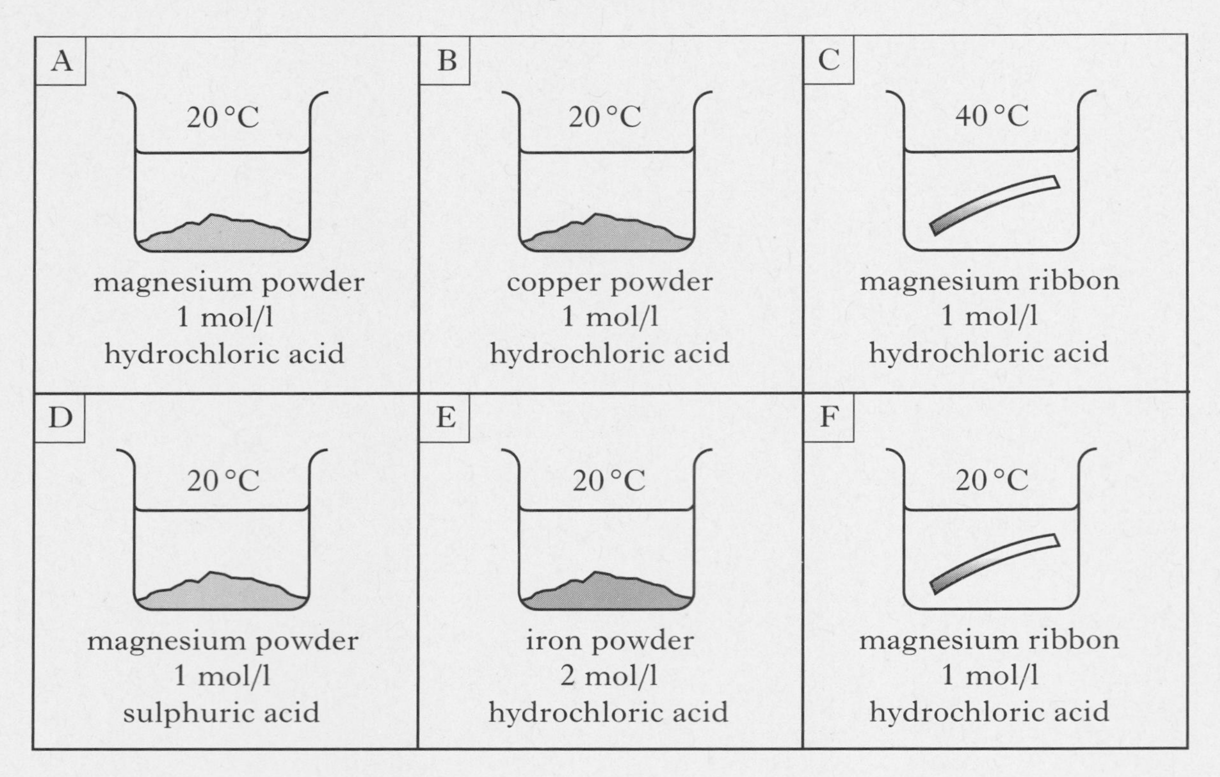


(*a*) Identify the **two** experiments which could be compared to show the effect of particle size on reaction rate.

| A | B | C |
|---|---|---|
| D | E | F |

**1**

(*b*) Identify the experiment in which **no** reaction would take place.

| A | B | C |
|---|---|---|
| D | E | F |

**1**

**(2)**

DO NOT WRITE IN THIS MARGIN

*Marks* KU PS

**3.** The grid shows the structural formulae of some hydrocarbons.

| A | B | C |
|---|---|---|
| H H<br>H–C—C–H<br>H H | H H<br>C<br>H H<br>C—C<br>H H | H H<br>C = C<br>H H |
| **D** | **E** | **F** |
| $CH_3$ H<br>C = C<br>H H | H $CH_3$ H<br>H–C–C—C–H<br>H $CH_3$ H | H H<br>H–C–C–H<br>H–C–C–H<br>H H |

(*a*) Identify the **two** hydrocarbons which can polymerise.

| A | B | C |
|---|---|---|
| D | E | F |

**1**

(*b*) Identify the **two** hydrocarbons with the general formula $C_nH_{2n}$ which do **not** decolourise bromine solution quickly.

| A | B | C |
|---|---|---|
| D | E | F |

**1**

**(2)**

**[Turn over**

DO NOT WRITE IN THIS MARGIN

*Marks* KU PS

**4.** The grid shows the names of some oxides.

| A silicon dioxide | B carbon dioxide | C sodium oxide |
|---|---|---|
| D iron oxide | E sulphur dioxide | F copper oxide |

(*a*) Identify the **two** oxides which contain transition metals.

You may wish to use the data booklet to help you.

| A | B | C |
|---|---|---|
| D | E | F |

1

(*b*) Identify the oxide which reacts with water in the atmosphere to produce acid rain.

| A | B | C |
|---|---|---|
| D | E | F |

1

(*c*) Identify the oxide which, when added to water, produces a solution with a greater concentration of hydroxide ions ($OH^-$) than hydrogen ions ($H^+$).

| A | B | C |
|---|---|---|
| D | E | F |

1

**(3)**

DO NOT WRITE IN THIS MARGIN

*Marks* KU PS

**5.** There are different types of chemical reactions.

| | |
|---|---|
| A | redox |
| B | precipitation |
| C | combustion |
| D | neutralisation |
| E | displacement |

(*a*) Identify the type of chemical reaction taking place when dilute hydrochloric acid reacts with a carbonate.

| |
|---|
| A |
| B |
| C |
| D |
| E |

**1**

(*b*) Identify the **two** types of chemical reaction represented by the following equation.

$$2Zn(s) + O_2(g) \longrightarrow 2ZnO(s)$$

| |
|---|
| A |
| B |
| C |
| D |
| E |

**2**

**(3)**

**[Turn over**

DO NOT WRITE IN THIS MARGIN

*Marks* KU PS

**6.** Lemonade can be made by dissolving sugar, lemon and carbon dioxide in water.

| | |
|---|---|
| A | sugar |
| B | lemon |
| C | carbon dioxide |
| D | water |

Identify the solvent used to make lemonade.

| |
|---|
| A |
| B |
| C |
| D |

**(1)**

DO NOT WRITE IN THIS MARGIN

Marks KU PS

**7.** The grid contains the names of some carbohydrates.

| | |
|---|---|
| A | fructose |
| B | glucose |
| C | maltose |
| D | sucrose |
| E | starch |

(*a*) Galactose is a monosaccharide found in dairy products.

Identify the **two** isomers of galactose.

| |
|---|
| A |
| B |
| C |
| D |
| E |

**1**

(*b*) Identify the carbohydrate which is a condensation polymer.

| |
|---|
| A |
| B |
| C |
| D |
| E |

**1**

**(2)**

**[Turn over**

DO NOT WRITE IN THIS MARGIN

Marks KU PS

8. A student made some statements about acids.

| | |
|---|---|
| A | Acid rain will have no effect on iron structures. |
| B | A base is a substance which can neutralise an alkali. |
| C | Treatment of acid indigestion is an example of neutralisation. |
| D | In a neutralisation reaction the pH of the acid will fall towards 7. |
| E | When dilute nitric acid reacts with potassium hydroxide solution, the salt potassium nitrate is produced. |

Identify the **two** correct statements.

| |
|---|
| A |
| B |
| C |
| D |
| E |

(2)

DO NOT WRITE IN THIS MARGIN

*Marks* KU PS

**9.** Coffee manufacturers have produced a self-heating can of coffee.

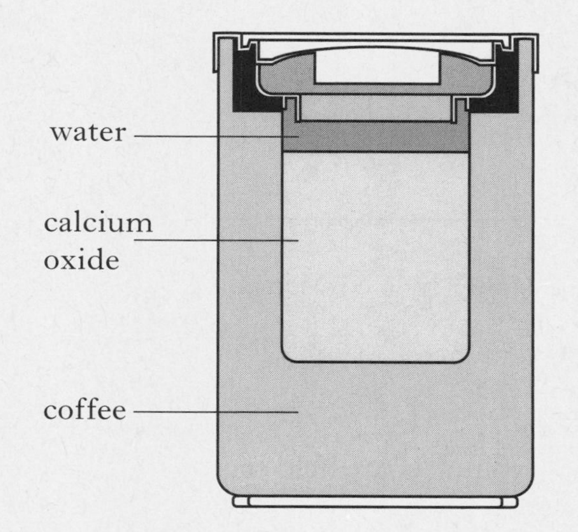

In the centre of the can calcium oxide reacts with water, releasing heat energy.

The equation for the reaction is:

$$CaO(s) + H_2O(\ell) \longrightarrow Ca(OH)_2(aq)$$

| | |
|---|---|
| A | Calcium oxide is insoluble. |
| B | The reaction is exothermic. |
| C | The reaction produces an acidic solution. |
| D | The temperature of the coffee goes down. |
| E | 0·1 moles of calcium oxide reacts with water producing 0·1 moles of calcium hydroxide. |

Identify the **two** correct statements.

| |
|---|
| A |
| B |
| C |
| D |
| E |

**(2)**

DO NOT WRITE IN THIS MARGIN

*Marks* KU PS

## PART 2

**A total of 40 marks is available in this part of the paper.**

**10.** Hydrogen reacts with other elements to form molecules such as hydrogen fluoride and hydrogen chloride.

(*a*) Name the family to which fluorine and chlorine belong.

_______________________________________ **1**

(*b*) The atoms in these molecules are held together by a covalent bond.

Circle the correct words to complete the sentence.

A covalent bond forms when two {positive / negative / neutral} nuclei are held together by their common attraction for a shared pair of {protons / neutrons / electrons}. **1**

(*c*) The table gives information about some molecules.

| Molecule H–X | Size of X/pm | Energy to break bond kJ/mol |
|---|---|---|
| H–F | 71 | 569 |
| H–Cl | 99 | 428 |
| H–Br | 114 | 362 |
| H–I | 133 | 295 |

Describe how the size of element **X** affects the energy needed to break the bond in the molecule.

_______________________________________

_______________________________________ **1**

**(3)**

*Marks* KU PS

**11.** Crude oil can be transported to a refinery through a steel pipeline.

(*a*) If the pipeline is not protected the iron will rust.

Name the **ion** formed from water and oxygen, when they accept electrons during rusting.

______________________ **1**

(*b*) Some parts of the pipeline are under the sea.

What effect would seawater have on the rate of rusting?

______________________________________________

______________________________________________ **1**

(*c*) Magnesium can be attached to the steel pipeline to prevent rusting.

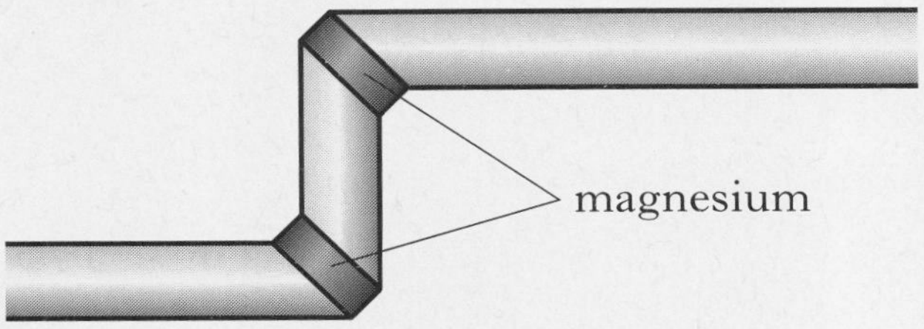

magnesium

What name is given to the **type** of protection provided by the magnesium?

______________________ **1**

**(3)**

**[Turn over**

DO NOT WRITE IN THIS MARGIN

*Marks* KU PS

**12.** Airbags in cars are designed to prevent injuries in car crashes.

They contain sodium azide ($NaN_3$) which produces nitrogen gas on impact.

The nitrogen inflates the airbag very quickly.

(*a*) The table gives information on the volume of nitrogen gas produced.

| Time/microseconds | Volume of nitrogen gas produced/litres |
|---|---|
| 0 | 0 |
| 5 | 46 |
| 10 | 64 |
| 15 | 74 |
| 20 | 82 |
| 25 | 88 |
| 30 | 88 |

(i) Draw a line graph of the results.

*Use appropriate scales to fill most of the graph paper.*

(Additional graph paper, if required, will be found on page 28.)

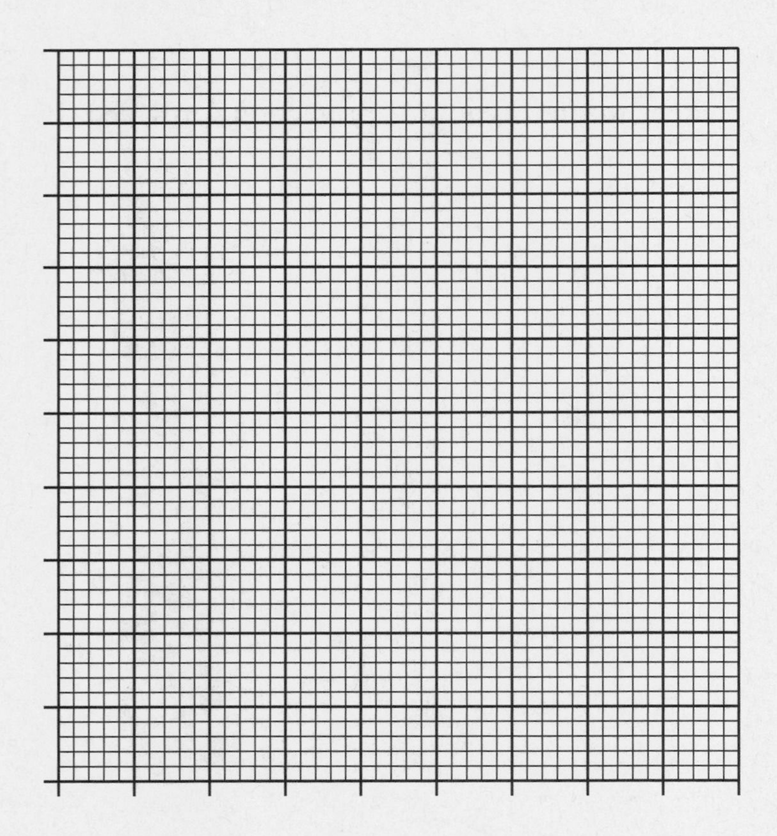

**2**

(ii) Using your graph, predict the time taken to produce 70 litres of nitrogen gas.

________________ microseconds **1**

DO NOT WRITE IN THIS MARGIN

*Marks* KU PS

**12. (continued)**

(*b*) The equation for the production of nitrogen gas is:

$$NaN_3(s) \longrightarrow N_2(g) + \quad Na(s).$$

Balance the equation above. **1**

(*c*) Nitrogen is a non-toxic gas.

Suggest another property of nitrogen which makes it a suitable gas for use in airbags.

______________________________________________________________

______________________________________________________________ **1**

**(5)**

**[Turn over**

DO NOT WRITE IN THIS MARGIN

*Marks* KU PS

**13.** Copper chloride solution can be broken up into its elements by passing electricity through it.

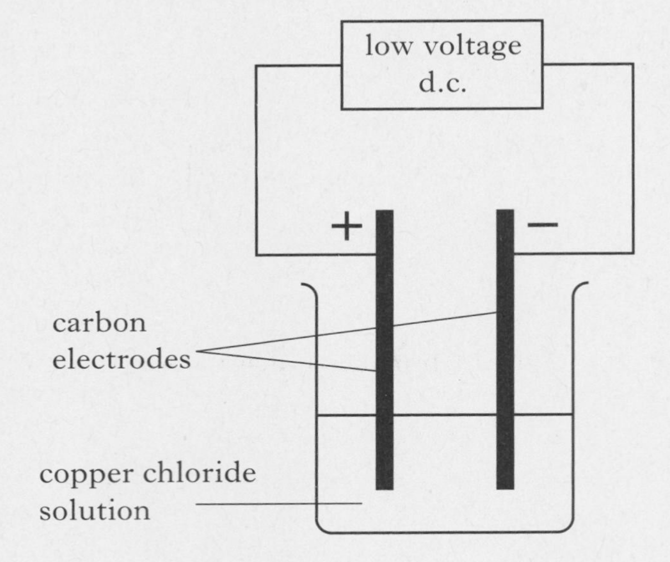


(*a*) Carbon is unreactive and insoluble in water.

Give another reason why it is suitable for use as an electrode.

________________________________________ **1**

(*b*) Chlorine gas is released at the positive electrode.

Write an ion-electron equation for the formation of chlorine.

You may wish to use the data booklet to help you.

________________________________________ **1**

(*c*) Why do ionic compounds, like copper chloride, conduct electricity when in solution?

________________________________________

________________________________________ **1**

**(3)**

DO NOT WRITE IN THIS MARGIN

*Marks* KU PS

**14.** A fizzy drink "Fizz Alive" contains a carbohydrate.

(*a*) Name all the elements found in a carbohydrate.

________________________________________ **1**

(*b*) A student carried out an investigation to find out which carbohydrate was present in "Fizz Alive".

**Test 1**

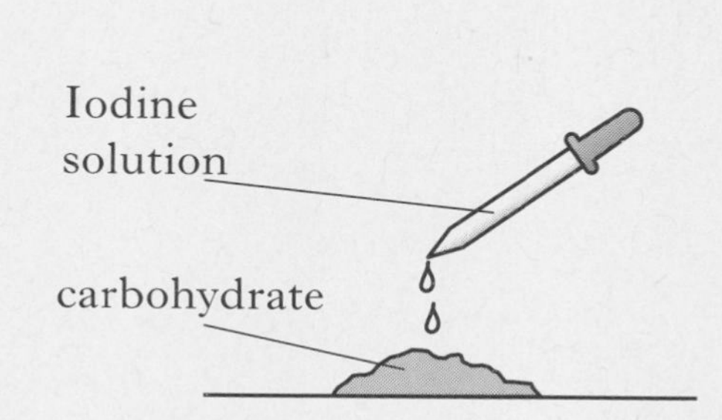


**Test 2**

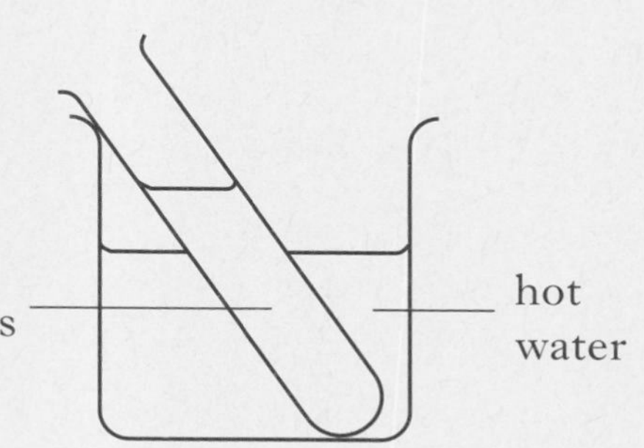


The results are shown in the table.

| Test | Result |
|---|---|
| Iodine solution | stays brown |
| Benedict's solution | stays blue |

Name the carbohydrate present in "Fizz Alive".

____________________ **1**

(*c*) A 330 $cm^3$ can of "Fizz Alive" has a carbohydrate concentration of 0·01 mol/l.

Calculate the number of moles of carbohydrate in the can of "Fizz Alive".

__________ mol **1**

**(3)**

**[Turn over**

DO NOT WRITE IN THIS MARGIN

*Marks* | KU | PS

**15.** The diagram represents the structure of an atom.

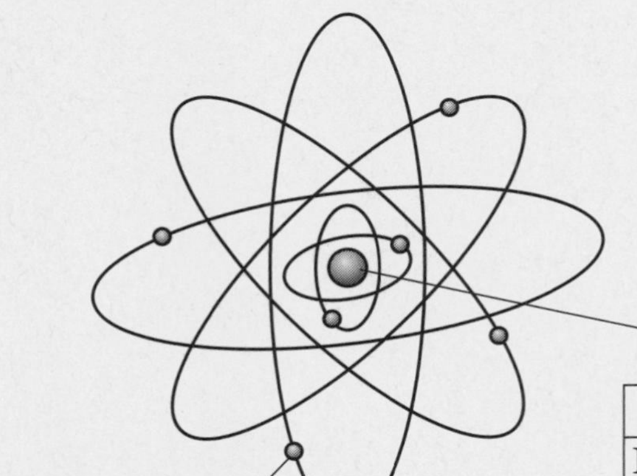

| THE NUCLEUS | |
|---|---|
| Name of Particle | Relative mass |
| PROTON | (i) |
| NEUTRON | 1 |

| OUTSIDE THE NUCLEUS | |
|---|---|
| Name of Particle | Relative mass |
| (ii) | 0 |

(*a*) Fill in the missing information for:

(i) ______________________________

(ii) ______________________________ **1**

DO NOT WRITE IN THIS MARGIN

*Marks* KU PS

**15. (continued)**

(*b*) The element uranium has unstable atoms.

These atoms give out radiation and a new element is formed.

$$^{238}_{92}U \longrightarrow ^{234}_{90}Th + ^{4}_{2}\alpha$$

radiation

(i) Complete the table to show the number of each type of particle in $^{234}_{90}Th$.

| Particle | Number |
|---|---|
| proton | |
| neutron | |

1

(ii) Radon is another element which gives out radiation.

$$^{222}_{86}Rn \longrightarrow \mathbf{X} + ^{4}_{2}\alpha$$

radiation

State the **atomic number** of element **X**.

____________________

1

**(3)**

**[Turn over**

DO NOT WRITE IN THIS MARGIN

Marks | KU | PS

**16.** Anglesite is an ore containing lead(II) sulphate, $PbSO_4$.

(*a*) Calculate the percentage by mass of lead in anglesite.

__________% **2**

(*b*) Most metals are found combined in the Earth's crust and have to be extracted from their ores.

Place the following metals in the correct space in the table.

**lead** **aluminium**

You may wish to use the data booklet to help you.

| Metal | Method of extraction |
|---|---|
| | electrolysis of molten compound |
| | using heat and carbon |

**1**

(*c*) Metal **X** can be extracted from its ore by heat alone.

What does this indicate about the reactivity of **X** compared to both lead and aluminium?

______________________________________

______________________________________ **1**

(*d*) When a metal is extracted from its ore, metal ions are changed to metal atoms.

Name this **type** of chemical reaction.

________________________ **1**

**(5)**

DO NOT WRITE IN THIS MARGIN

*Marks* KU PS

**17.** A student added strips of magnesium to solutions of other metals.

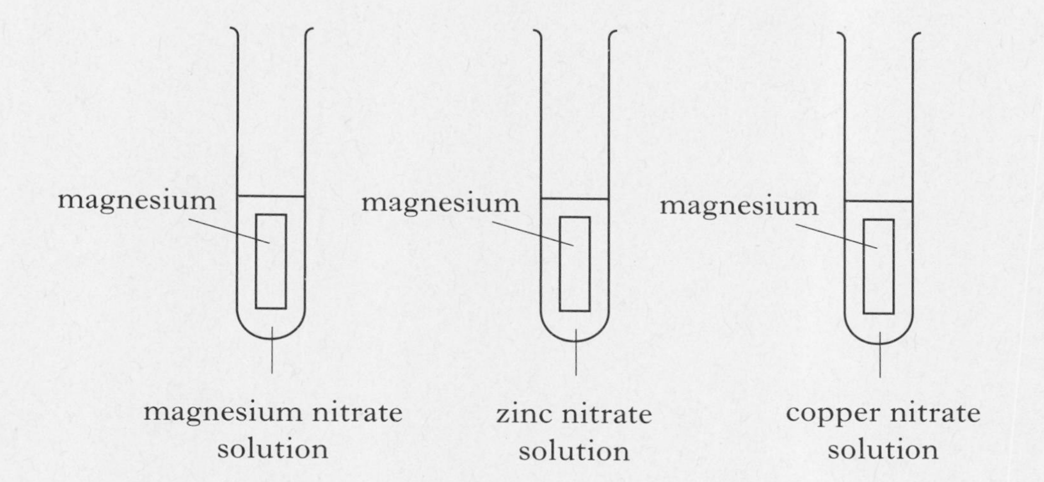


The results are shown in the table.

| Metal \ Solution | **magnesium nitrate** | **zinc nitrate** | **copper nitrate** |
|---|---|---|---|
| **magnesium** | (i) | (ii) | reaction occurred |

(*a*) In the table, fill in the missing information at (i) and (ii) to show whether or not a chemical reaction has occurred.

You may wish to use the data booklet to help you. **1**

(*b*) The equation for the reaction between magnesium and copper nitrate is:

$$Mg(s) + Cu^{2+}(aq) + 2NO_3^-(aq) \longrightarrow Mg^{2+}(aq) + 2NO_3^-(aq) + Cu(s).$$

(i) Circle the spectator ion in the above equation. **1**

(ii) What technique could be used to remove copper from the mixture?

______________________________

______________________________ **1**

**(3)**

**[Turn over**

DO NOT WRITE IN THIS MARGIN

Marks | KU | PS

**18.** Nitrogen is essential for healthy plant growth.

Nitrogen from the atmosphere can be fixed in a number of ways.

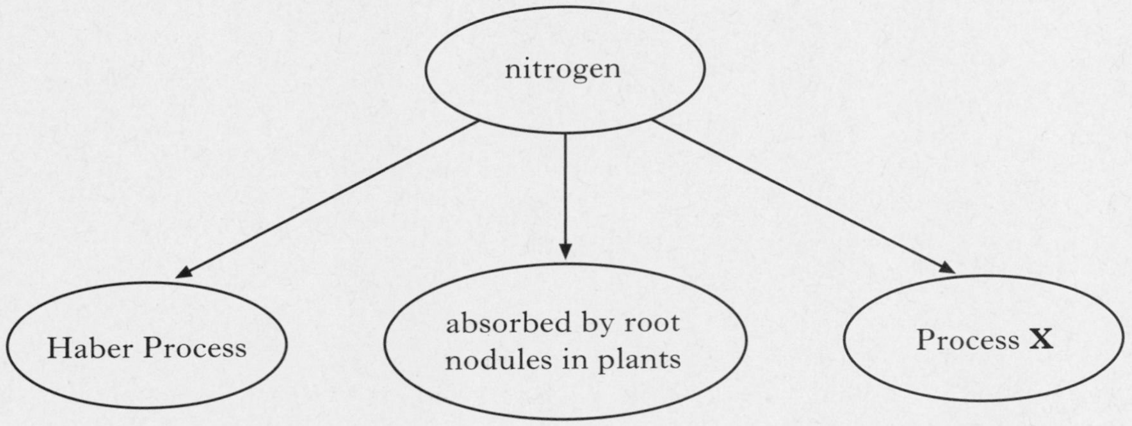

nitrogen → Haber Process
nitrogen → absorbed by root nodules in plants
nitrogen → Process **X**

(*a*) **X** is a natural process which takes place in the atmosphere, producing nitrogen dioxide gas.

What provides the energy for this process?

________________________________________ **1**

(*b*) What is present in the root nodules of some plants which convert nitrogen from the atmosphere into nitrogen compounds?

________________________________________

________________________________________ **1**

(*c*) The Haber Process is the industrial method of converting nitrogen into a nitrogen compound.

Name the nitrogen compound produced.

____________________ **1**

DO NOT WRITE IN THIS MARGIN

*Marks* | KU | PS

**18. (continued)**

(*d*) The nitrogen compound produced in the Haber Process dissolves in water.

The graph shows the solubility of the nitrogen compound at different temperatures.

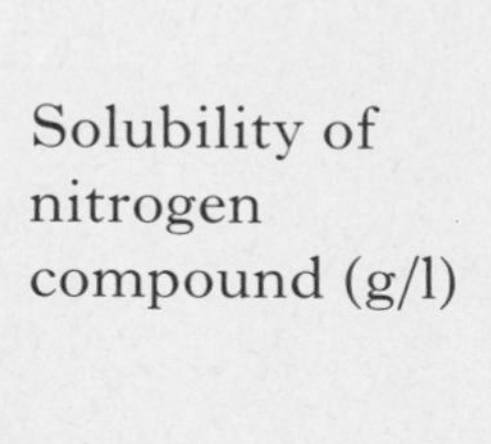


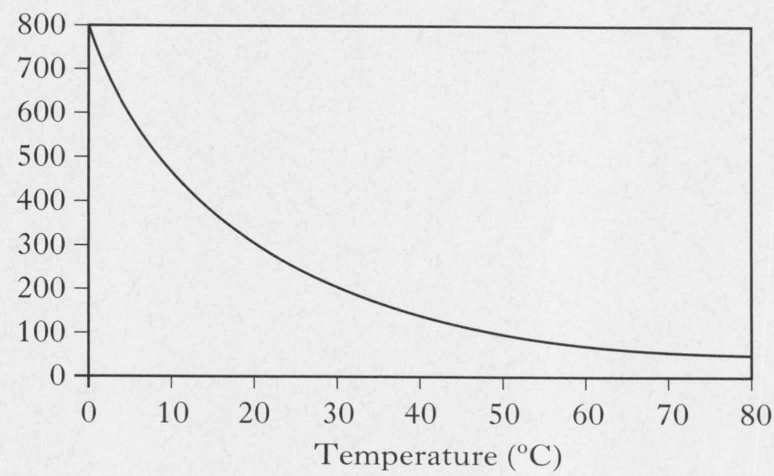


Write a general statement describing the effect of temperature on the solubility of the nitrogen compound.

______________________________________________________________

______________________________________________________________ **1**

**(4)**

**[Turn over**

DO NOT WRITE IN THIS MARGIN

*Marks* KU PS

**19.** The octane number indicates how efficiently a fuel burns.

| Alkane | Molecular Formula | Full Structural Formula | Octane Number |
|---|---|---|---|
| 2-methylbutane | $C_5H_{12}$ | H<br>H–C–H<br>H H | H<br>H–C–C–C–C–H<br>H H H H | 93 |
| 2-methylpentane | $C_6H_{14}$ | H<br>H–C–H<br>H H H | H<br>H–C–C–C–C–C–H<br>H H H H H | 71 |
| 2-methylhexane | $C_7H_{16}$ | | 47 |
| 2-methylheptane | $C_8H_{18}$ | H<br>H–C–H<br>H H H H H | H<br>H–C–C–C–C–C–C–C–H<br>H H H H H H H | |
| 2-methyloctane | $C_9H_{20}$ | H<br>H–C–H<br>H H H H H H | H<br>H–C–C–C–C–C–C–C–C–H<br>H H H H H H H H | 2 |

(*a*) Draw the **full** structural formula for 2-methylhexane.

1

DO NOT WRITE IN THIS MARGIN

Marks KU PS

**19. (continued)**

(*b*) 2-methylpentane and hexane have the same molecular formula ($C_6H_{14}$), but different structural formulae.

What term is used to describe this pair of alkanes?

________________________ **1**

(*c*) Using information in the table, predict the octane number for 2-methylheptane.

________________________ **1**

**(3)**

**[Turn over**

DO NOT WRITE IN THIS MARGIN

*Marks* KU PS

**20.** Molten iron is used to join steel railway lines together.

Molten iron is produced when aluminium reacts with iron oxide.

The equation for the reaction is:

$$2Al + Fe_2O_3 \longrightarrow 2Fe + Al_2O_3$$

(*a*) Calculate the mass of iron produced from 40 grams of iron oxide.

__________ g **2**

(*b*) The formula for iron oxide is $Fe_2O_3$.

What is the charge on this iron ion?

__________ **1**

DO NOT WRITE IN THIS MARGIN

Marks KU PS

**20. (continued)**

(*c*) Iron can also be produced from iron ore, $Fe_2O_3$, in a blast furnace.

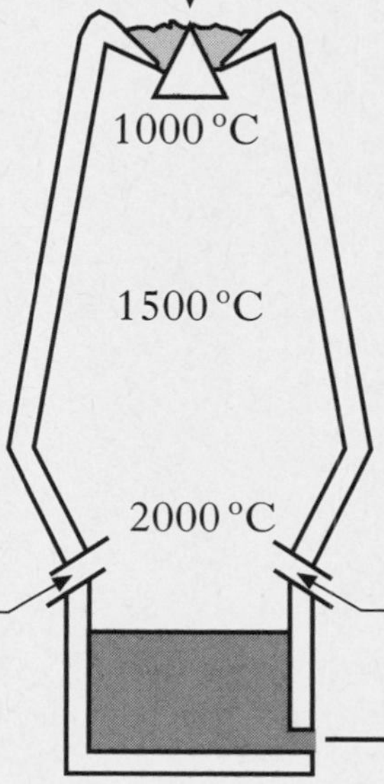

iron ore, carbon and limestone

1000 °C

1500 °C

2000 °C

air

air

molten iron

The main reactions taking place are:

$$C(s) + O_2(g) \longrightarrow CO_2(g)$$

$$CO_2(g) + C(s) \longrightarrow 2CO(g)$$

$$Fe_2O_3(s) + 3CO(g) \longrightarrow 2Fe(\ell) + 3CO_2(g)$$

(i) When air is blown into the furnace the temperature rises.

Suggest another reason why **air** is blown into the furnace.

______________________________________________

______________________________________________ **1**

(ii) Explain why the temperature at the bottom of the blast furnace should **not** drop below 1535 °C.

You may wish to use the data booklet to help you.

______________________________________________

______________________________________________ **1**

**(5)**

[*END OF QUESTION PAPER*]

DO NOT WRITE IN THIS MARGIN

| KU | PS |
|---|---|

## ADDITIONAL SPACE FOR ANSWERS

ADDITIONAL GRAPH PAPER FOR QUESTION 12(*a*)(i)

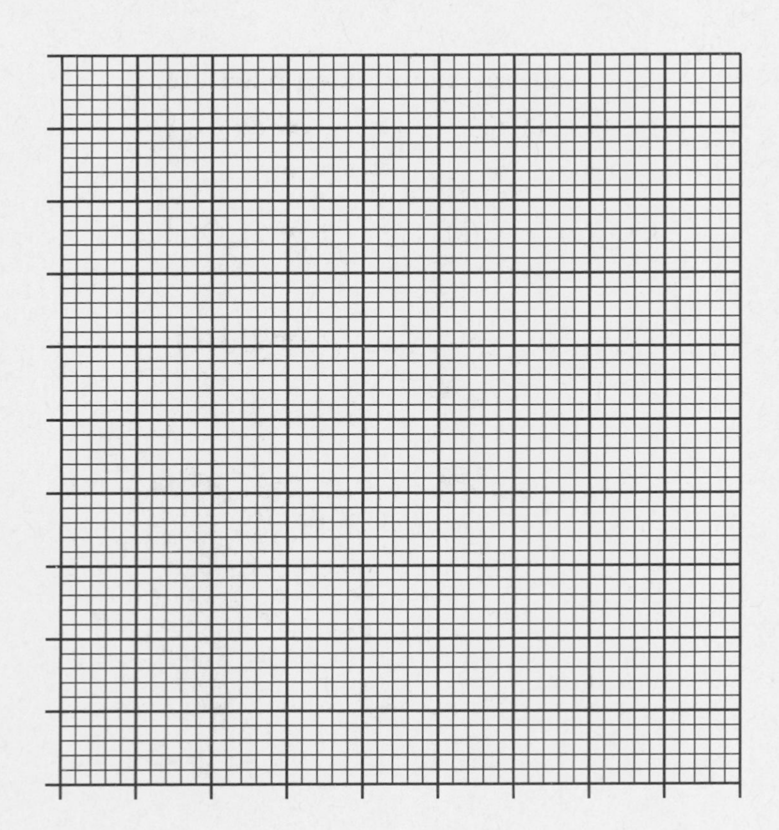

STANDARD GRADE | CREDIT

# 2009

[BLANK PAGE]

FOR OFFICIAL USE

| | | | | | |
|---|---|---|---|---|---|
| | | | | | |

C

| | KU | PS |
|---|---|---|
| Total Marks | | |

# 0500/402

NATIONAL QUALIFICATIONS 2009

MONDAY, 11 MAY
10.50 AM – 12.20 PM

# CHEMISTRY
## STANDARD GRADE
### Credit Level

**Fill in these boxes and read what is printed below.**

Full name of centre

Town

Forename(s)

Surname

Date of birth
Day Month Year

Scottish candidate number

Number of seat

1 All questions should be attempted.

2 Necessary data will be found in the Data Booklet provided for Chemistry at Standard Grade and Intermediate 2.

3 The questions may be answered in any order but all answers are to be written in this answer book, and must be written clearly and legibly in ink.

4 Rough work, if any should be necessary, as well as the fair copy, is to be written in this book.

Rough work should be scored through when the fair copy has been written.

5 Additional space for answers and rough work will be found at the end of the book.

6 The size of the space provided for an answer should not be taken as an indication of how much to write. It is not necessary to use all the space.

7 Before leaving the examination room you must give this book to the invigilator. If you do not, you may lose all the marks for this paper.

SA 0500/402 6/24620

## PART 1

**In Questions 1 to 8 of this part of the paper, an answer is given by circling the appropriate letter (or letters) in the answer grid provided.**

**In some questions, two letters are required for full marks.**

**If more than the correct number of answers is given, marks will be deducted.**

**A total of 20 marks is available in this part of the paper.**

### SAMPLE QUESTION

| A | B | C |
|---|---|---|
| $CH_4$ | $H_2$ | $CO_2$ |
| D | E | F |
| $CO$ | $C_2H_5OH$ | $C$ |

(*a*) Identify the hydrocarbon.

| (A) | B | C |
|---|---|---|
| D | E | F |

The one correct answer to part (*a*) is A. This should be circled.

(*b*) Identify the **two** elements.

| A | (B) | C |
|---|---|---|
| D | E | (F) |

As indicated in this question, there are **two** correct answers to part (*b*). These are B and F. Both answers are circled.

If, after you have recorded your answer, you decide that you have made an error and wish to make a change, you should cancel the original answer and circle the answer you now consider to be correct. Thus, in part (*a*), if you want to change an answer A to an answer D, your answer sheet would look like this:

| ~~(A)~~ | B | C |
|---|---|---|
| (D) | E | F |

If you want to change back to an answer which has already been scored out, you should enter a tick (✓) in the box of the answer of your choice, thus:

| ✓ ~~(A)~~ | B | C |
|---|---|---|
| ~~(D)~~ | E | F |

DO NOT WRITE IN THIS MARGIN

Marks KU PS

**1.** Many solutions are used for chemical tests.

| A Benedict's reagent | B lime water | C bromine solution |
|---|---|---|
| D pH indicator | E iodine solution | F ferroxyl indicator |

(*a*) Identify the solution which could be used to test for maltose.

| A | B | C |
|---|---|---|
| D | E | F |

1

(*b*) Identify the solution which is used to test for $Fe^{2+}(aq)$.

| A | B | C |
|---|---|---|
| D | E | F |

1

**(2)**

**[Turn over**

DO NOT WRITE IN THIS MARGIN

Marks | KU | PS

**2.** Many chemical compounds contain ions.

| A | B | C |
|---|---|---|
| strontium chloride | lithium oxide | calcium oxide |
| **D** | **E** | **F** |
| barium fluoride | sodium fluoride | potassium chloride |

(*a*) Identify the compound which produces a green flame colour.

You may wish to use the data booklet to help you.

| A | B | C |
|---|---|---|
| D | E | F |

**1**

(*b*) Identify the compound in which **both** ions have the same electron arrangement as argon.

| A | B | C |
|---|---|---|
| D | E | F |

**1**

**(2)**

DO NOT WRITE IN THIS MARGIN

Marks KU PS

3. The table contains information about some substances.

| Substance | Melting point/°C | Boiling point/°C | Conducts as | |
|---|---|---|---|---|
| | | | a solid | a liquid |
| A | 639 | 3228 | yes | yes |
| B | 2967 | 3273 | no | no |
| C | 159 | 211 | no | no |
| D | 1402 | 2497 | no | yes |
| E | 27 | 677 | yes | yes |

(*a*) Identify the substance which exists as a covalent network.

| A |
|---|
| B |
| C |
| D |
| E |

1

(*b*) Identify the substance which could be calcium fluoride.

| A |
|---|
| B |
| C |
| D |
| E |

1

(2)

[Turn over

DO NOT WRITE IN THIS MARGIN

Marks | KU | PS

**4.** The grid shows the names of some ionic compounds.

| A aluminium bromide | B sodium chloride | C potassium hydroxide |
|---|---|---|
| D sodium sulphate | E potassium bromide | F calcium chloride |

(*a*) Identify the base.

| A | B | C |
|---|---|---|
| D | E | F |

**1**

(*b*) Identify the **two** compounds whose solutions would form a precipitate when mixed.

You may wish to use the data booklet to help you.

| A | B | C |
|---|---|---|
| D | E | F |

**1**

(*c*) Identify the compound with a formula of the type $\mathbf{XY_2}$, where **X** is a metal.

| A | B | C |
|---|---|---|
| D | E | F |

**1**

**(3)**

DO NOT WRITE IN THIS MARGIN

Marks | KU | PS

5. The names of some hydrocarbons are shown in the grid.

| A ethane | B pentene | C cyclohexane |
|---|---|---|
| D pentane | E cyclopentane | F propene |

(*a*) Identify the **two** isomers.

| A | B | C |
|---|---|---|
| D | E | F |

1

(*b*) Identify the hydrocarbon with the highest boiling point.

You may wish to use the data booklet to help you.

| A | B | C |
|---|---|---|
| D | E | F |

1

(*c*) Identify the **two** hydrocarbons which can take part in an addition reaction with hydrogen.

| A | B | C |
|---|---|---|
| D | E | F |

1

**(3)**

**[Turn over**

DO NOT WRITE IN THIS MARGIN

Marks | KU | PS

**6.** Reactions can be represented using chemical equations.

| | |
|---|---|
| A | $Fe^{2+}(aq) + 2e^{-} \rightarrow Fe(s)$ |
| B | $Fe^{2+}(aq) \rightarrow Fe^{3+}(aq) + e^{-}$ |
| C | $2H_2(g) + O_2(g) \rightarrow 2H_2O(g)$ |
| D | $2H_2O(\ell) + O_2(g) + 4e^{-} \rightarrow 4OH^{-}(aq)$ |
| E | $SO_2(g) + H_2O(\ell) \rightarrow 2H^{+}(aq) + SO_3^{2-}(aq)$ |

(*a*) Identify the equation which shows the formation of acid rain.

| |
|---|
| A |
| B |
| C |
| D |
| E |

1

(*b*) Identify the equation which represents a combustion reaction.

| |
|---|
| A |
| B |
| C |
| D |
| E |

1

(*c*) Identify the **two** equations which are involved in the corrosion of iron.

| |
|---|
| A |
| B |
| C |
| D |
| E |

2

**(4)**

DO NOT WRITE IN THIS MARGIN

*Marks* KU PS

**7.** The grid contains information about the particles found in atoms.

| A | B | C |
|---|---|---|
| relative mass = 1 | charge = zero | relative mass almost zero |
| D | E | F |
| charge = 1– | found outside the nucleus | charge = 1+ |

Identify the **two** terms which can be applied to protons.

| A | B | C |
|---|---|---|
| D | E | F |

**(2)**

**[Turn over**

DO NOT WRITE IN THIS MARGIN

*Marks* KU PS

**8.** The fractional distillation of crude oil was demonstrated to a class.

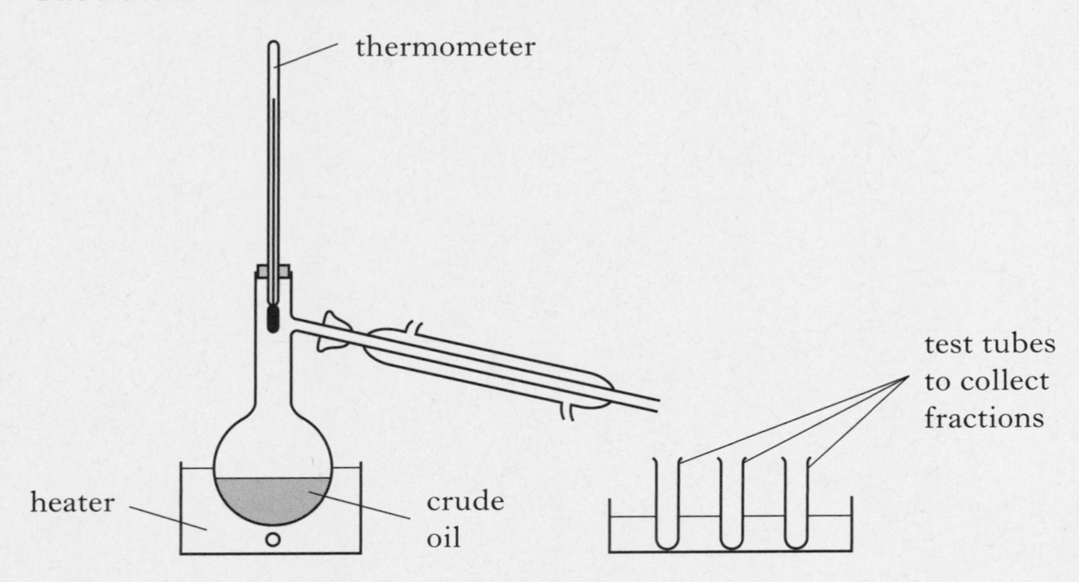


Six fractions were numbered in the order they were collected.

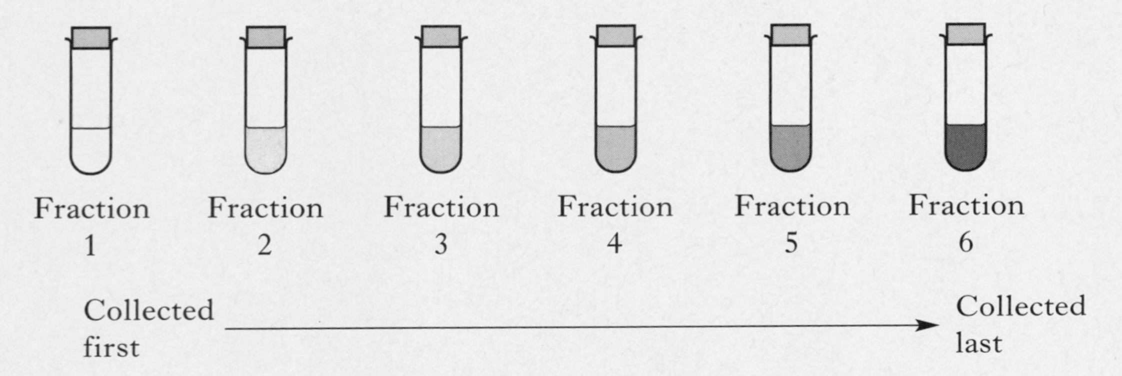


Identify the **two** correct statements.

| | |
|---|---|
| A | Fraction 6 evaporates most easily. |
| B | Fraction 5 is less viscous than fraction 4. |
| C | Fraction 2 is more flammable than fraction 3. |
| D | Fraction 1 has a lower boiling range than fraction 2. |
| E | The molecules in fraction 3 are larger than those in fraction 4. |

| |
|---|
| A |
| B |
| C |
| D |
| E |

**(2)**

**[Turn over for Part 2 on *Page twelve***

DO NOT WRITE IN THIS MARGIN

*Marks* KU PS

## PART 2

**A total of 40 marks is available in this part of the paper.**

**9.** There are three different types of neon atom.

| Type of atom | Number of protons | Number of neutrons |
|---|---|---|
| $^{20}_{10}Ne$ | | |
| $^{21}_{10}Ne$ | | |
| $^{22}_{10}Ne$ | | |

(*a*) Complete the table to show the number of protons and neutrons in each type of neon atom. **1**

(*b*) What term is used to describe these different types of neon atom?

________________________________________ **1**

(*c*) A natural sample of neon has an average atomic mass of 20·2.

What is the mass number of the most common type of atom in the sample of neon?

____________________ **1**

**(3)**

DO NOT WRITE IN THIS MARGIN

*Marks* KU PS

**10.** Aluminium metal can be produced by passing electricity through molten aluminium oxide.

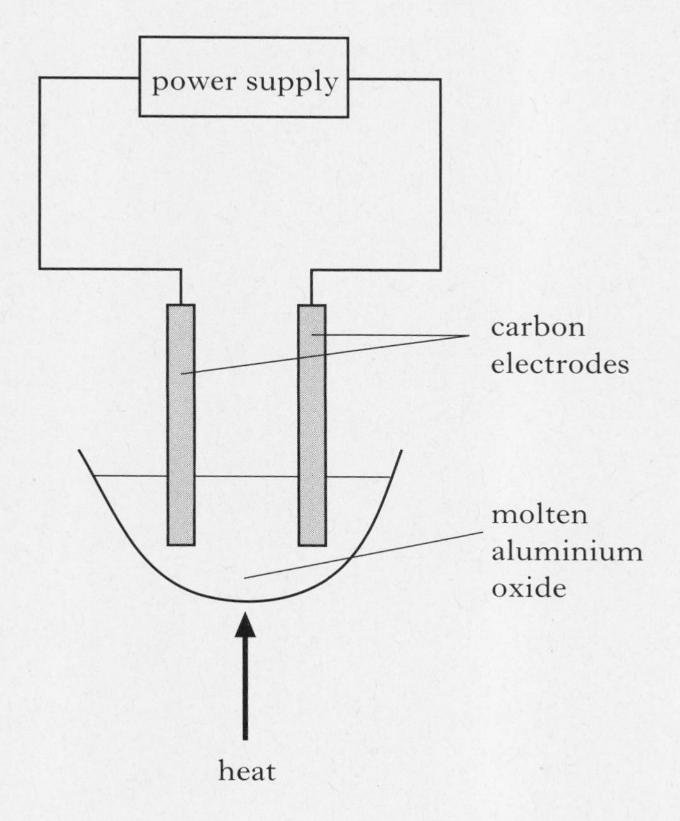


(*a*) Name this process.

______________________________ **1**

(*b*) Write the ionic formula for aluminium oxide.

**1**

(*c*) Why do ionic compounds, like aluminium oxide, conduct electricity when molten?

______________________________

______________________________ **1**

**(3)**

**[Turn over**

DO NOT WRITE IN THIS MARGIN

*Marks* KU PS

**11.** A student burned gas **X** and the products were passed through the apparatus shown.

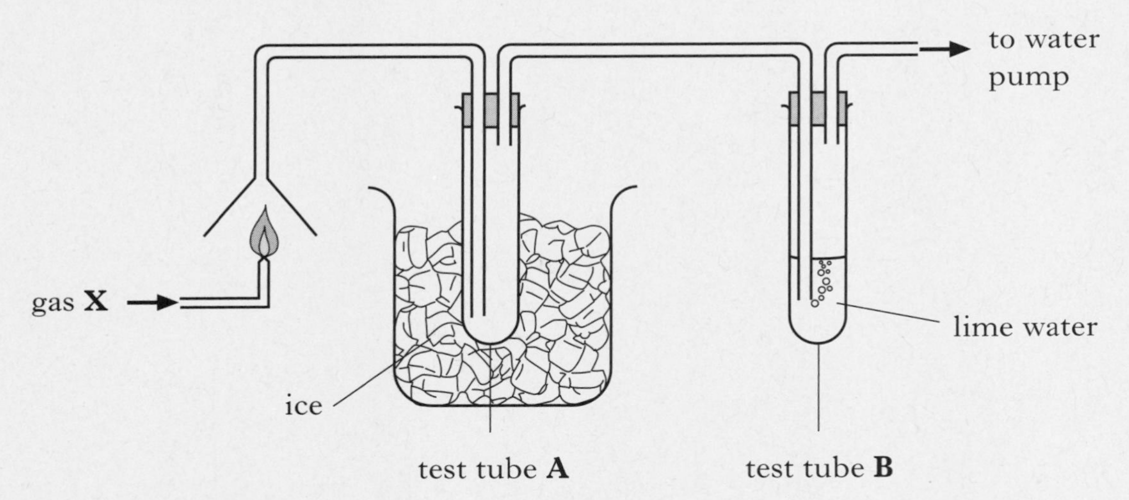


(*a*) The results are shown in the table.

| Observation in test tube A | Observation in test tube B |
|---|---|
| colourless liquid forms | lime water turns milky |

Using the information in the table, name two **elements** which **must** be present in gas **X**.

________________________________________ 1

(*b*) The experiment was repeated using hydrogen gas.

Complete the table showing the results which would have been obtained.

| Observation in test tube A | Observation in test tube B |
|---|---|
| | |

1

**(2)**

DO NOT WRITE IN THIS MARGIN

*Marks* KU PS

**12.** Hydrogen can form bonds with other elements.

The diagram shows the arrangement of outer electrons in a molecule of hydrogen chloride.

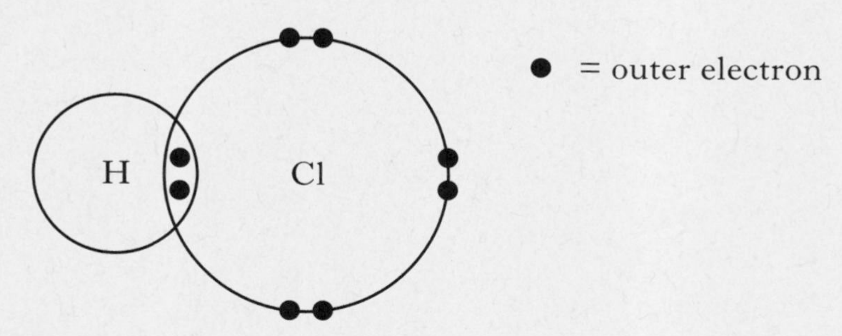


(*a*) What type of bonding is present in a hydrogen chloride molecule?

______________________________________________ **1**

(*b*) Draw a similar diagram, showing **all** outer electrons, to represent a molecule of phosphine, $PH_3$.

**1**

**(2)**

**[Turn over**

DO NOT WRITE IN THIS MARGIN

*Marks* KU PS

**13.** The apparatus below was used to investigate the reaction between lumps of calcium carbonate and dilute hydrochloric acid.

Excess acid was used to make sure all the calcium carbonate reacted.

A balance was used to measure the mass lost during the reaction.

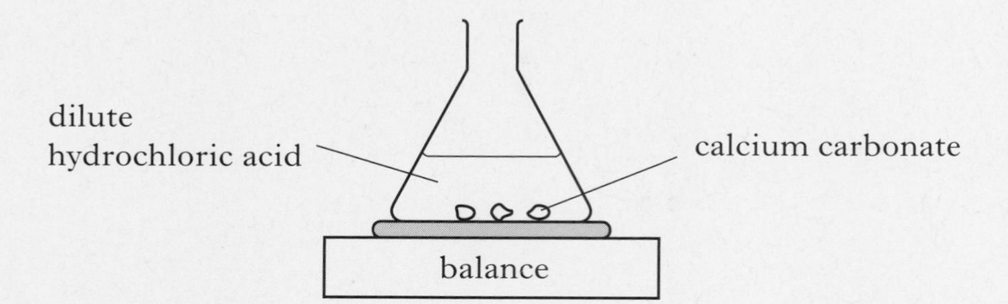


(*a*) Name the type of chemical reaction taking place when calcium carbonate reacts with dilute hydrochloric acid.

________________________________________ **1**

(*b*) The results are shown in the table.

| **Time/minutes** | 0 | 0·5 | 1·0 | 2·0 | 3·0 | 4·0 | 5·0 |
|---|---|---|---|---|---|---|---|
| **Mass lost/g** | 0 | 0·36 | 0·52 | 0·70 | 0·80 | 0·86 | 0·86 |

(i) Why is mass lost during the reaction?

________________________________________

________________________________________ **1**

DO NOT WRITE IN THIS MARGIN

*Marks* KU PS

**13. (*b*) (continued)**

(ii) Draw a line graph of the results.

*Use appropriate scales to fill most of the graph paper.*

(Additional graph paper, if required, will be found on page 26.)

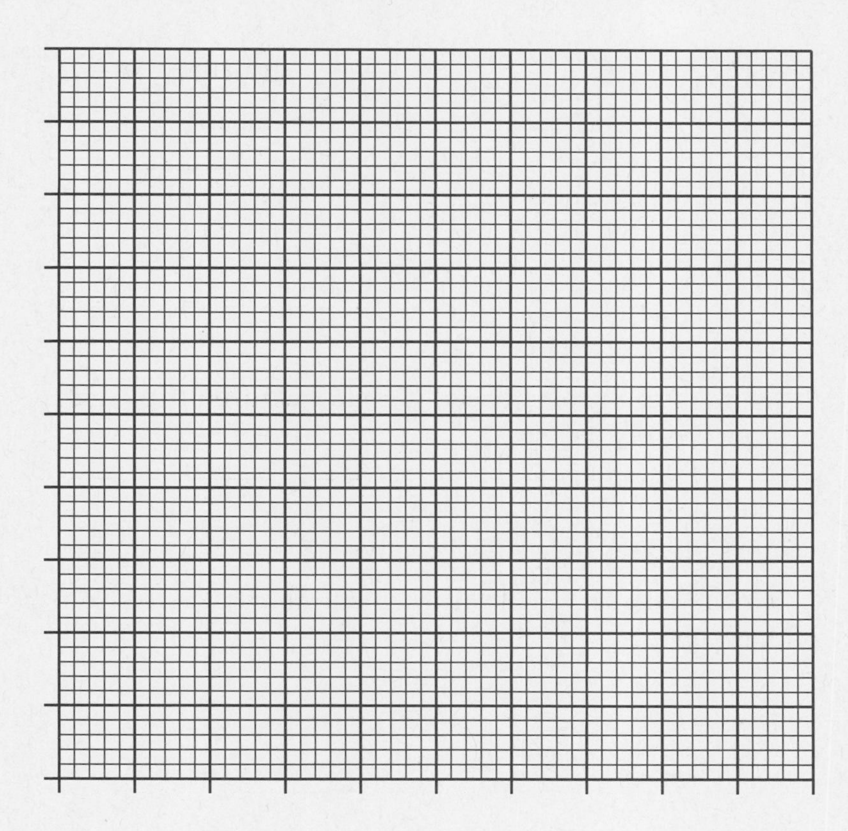

**2**

(*c*) The experiment was repeated using the same volume and concentration of acid. The same mass of calcium carbonate was used but **powder** instead of lumps.

Suggest how much mass would have been lost after three minutes.

__________ g **1**

**(5)**

**[Turn over**

DO NOT WRITE IN THIS MARGIN

*Marks* KU PS

**14.** (*a*) The flow diagram shows how ammonia is converted to nitric acid.

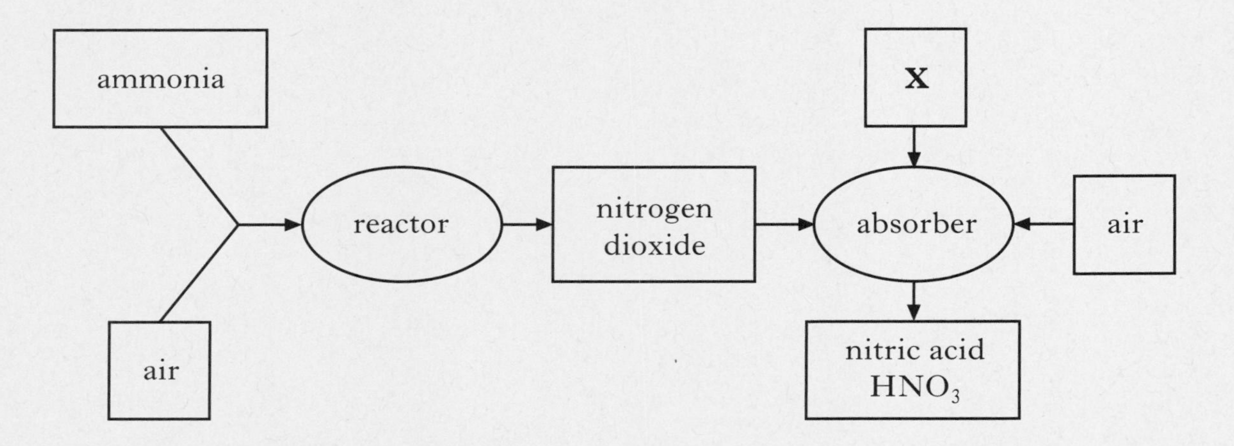


(i) Name the industrial process used to manufacture nitric acid.

______________________________ **1**

(ii) The reactor contains a platinum catalyst.

Why is it **not** necessary to continue heating the catalyst once the reaction has started?

______________________________

______________________________ **1**

(iii) Name substance **X**.

______________________________ **1**

(*b*) Ammonia and nitric acid react together to form ammonium nitrate, $NH_4NO_3$.

Calculate the percentage by mass of nitrogen in ammonium nitrate.

**Show your working clearly.**

________ % **2**

**(5)**

DO NOT WRITE IN THIS MARGIN

*Marks* KU PS

**15.** A student carried out some experiments with four metals and their oxides.

The results are shown in the table.

| Metal | Reaction with cold water | Reaction with dilute acid | Effect of heat on metal oxide |
|---|---|---|---|
| **W** | no reaction | no reaction | no reaction |
| **X** | no reaction | gas produced | no reaction |
| **Y** | gas produced | gas produced | no reaction |
| **Z** | no reaction | no reaction | metal produced |

(*a*) Place the four metals in order of reactivity (**most reactive first**).

_______________________________________________ **1**

(*b*) Name the gas produced when metal **Y** reacts with cold water.

_______________________________________________ **1**

(*c*) Suggest names for metals **Y** and **Z**.

metal **Y** ____________________ metal **Z** ____________________ **1**

**(3)**

**[Turn over**

DO NOT WRITE IN THIS MARGIN

Marks | KU | PS

**16.** The diagram shows the main stages in the making of malt whisky.

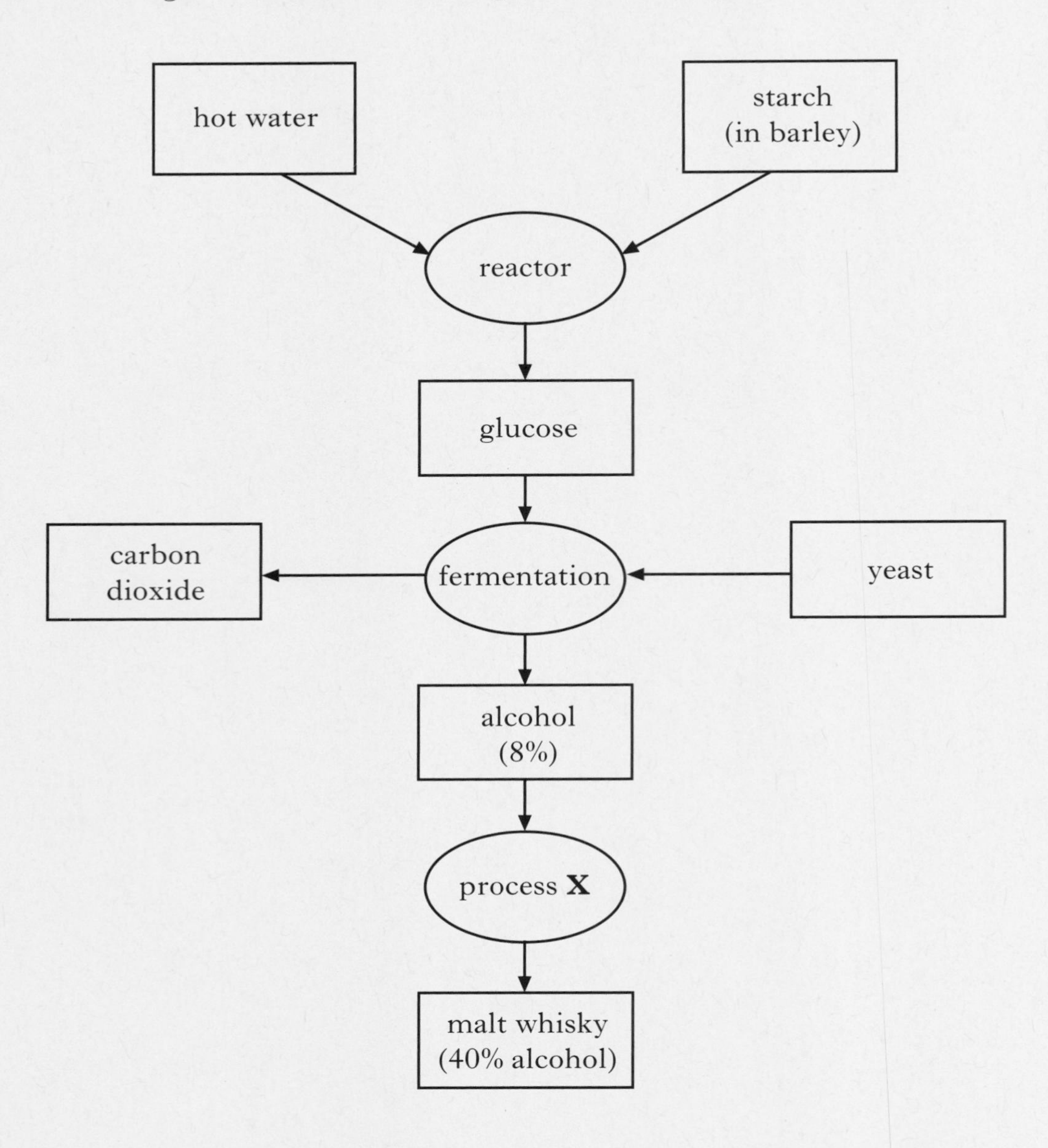


(*a*) Name the type of chemical reaction which takes place in the reactor.

________________________________________ **1**

DO NOT WRITE IN THIS MARGIN

Marks KU PS

**16. (continued)**

(*b*) The equation for the reaction taking place during fermentation is:

$$C_6H_{12}O_6 \longrightarrow C_2H_5OH + CO_2$$

Balance this equation. **1**

(*c*) What name is given to process **X**?

________________________________________ **1**

(*d*) Ethanol, $C_2H_5OH$, is the alcohol found in whisky.

A bottle of whisky contains 230 g of ethanol.

Calculate the number of moles of ethanol present in the whisky.

**Show your working clearly.**

____________ mol **2**

**(5)**

**[Turn over**

DO NOT WRITE IN THIS MARGIN

*Marks* KU PS

**17.** A student set up the cell shown.

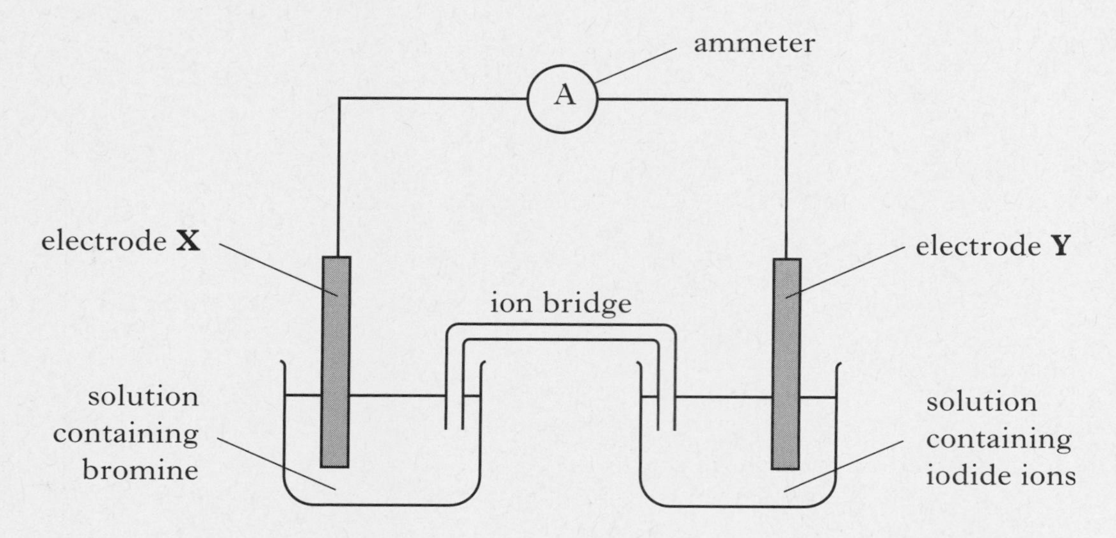


The reaction taking place at electrode **Y** is:

$$2I^-(aq) \longrightarrow I_2(s) + 2e^-$$

(*a*) Name the type of chemical reaction taking place at electrode **Y**.

______________________________ **1**

(*b*) **On the diagram**, clearly mark the path and direction of the electron flow. **1**

(*c*) Describe a test, including the result, which would show that iodine had formed at electrode **Y**.

______________________________

______________________________ **1**

(*d*) Write the ion-electron equation for the chemical reaction taking place at electrode **X**.

**1**

**(4)**

DO NOT WRITE IN THIS MARGIN

Marks | KU | PS

**18.** When superglue sets, a polymer is formed.

Part of the polymer structure is shown.

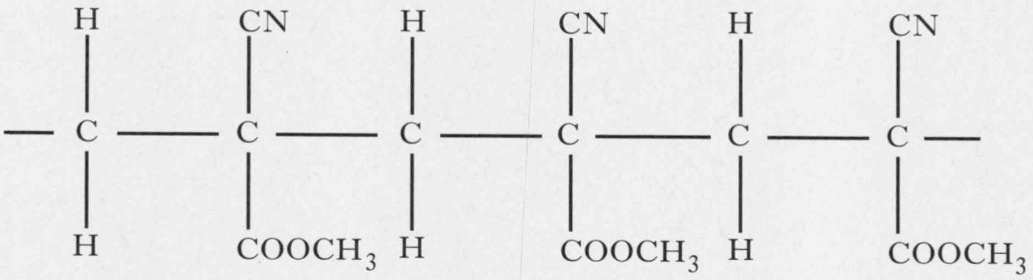

```
     H        CN       H        CN       H        CN
     |        |        |        |        |        |
  —  C  ———   C  ———   C  ———   C  ———   C  ———   C  —
     |        |        |        |        |        |
     H      COOCH3     H      COOCH3     H      COOCH3
```

(*a*) Draw the structure of the repeating unit in the superglue polymer. **1**

(*b*) The polymer shown above contains methyl groups ($CH_3$).

Another type of superglue, used to close cuts, has the methyl groups replaced by either butyl groups ($C_4H_9$) or octyl groups.

Complete the table to show the number of carbon and hydrogen atoms in an octyl group.

| | **Number of atoms** | |
|---|---|---|
| **Group** | **Carbon** | **Hydrogen** |
| methyl | 1 | 3 |
| butyl | 4 | 9 |
| octyl | | |

**1**

(*c*) Name a toxic gas given off when superglue burns.

________________________________________ **1**

**(3)**

**[Turn over**

DO NOT WRITE IN THIS MARGIN

Marks KU PS

**19.** (*a*) The table gives information about some members of the alkane family.

| **Name** | **Molecular formula** | **Boiling point/°C** |
|---|---|---|
| nonane | $C_9H_{20}$ | 151 |
| decane | $C_{10}H_{22}$ | 174 |
| undecane | $C_{11}H_{24}$ | 196 |
| dodecane | $C_{12}H_{26}$ | |

Predict the boiling point of dodecane.

________________ °C **1**

(*b*) What term is used to describe any family of compounds, like the alkanes, which have the same general formula and similar chemical properties?

________________________________________ **1**

(*c*) The equation for the burning of nonane is:

$$C_9H_{20} + 14O_2 \longrightarrow 9CO_2 + 10H_2O$$

Calculate the mass of water produced when 6·4 grams of nonane is burned.

**Show your working clearly.**

____________ g **2**

DO NOT WRITE IN THIS MARGIN

*Marks* KU PS

**19. (continued)**

(*d*) Alkanes can be prepared by the Kolbé synthesis.

$$CH_3COO^- + {}^-OOCCH_3 \longrightarrow CH_3CH_3 + 2CO_2 + 2e^-$$

ethanoate ions → ethane

Draw a structural formula for the alkane produced when propanoate ions are used instead of ethanoate ions.

$$CH_3CH_2COO^- + {}^-OOCCH_2CH_3 \longrightarrow$$

1

(5)

[*END OF QUESTION PAPER*]

DO NOT WRITE IN THIS MARGIN

| KU | PS |
|---|---|

## ADDITIONAL SPACE FOR ANSWERS

ADDITIONAL GRAPH PAPER FOR QUESTION 13(*b*)(ii)

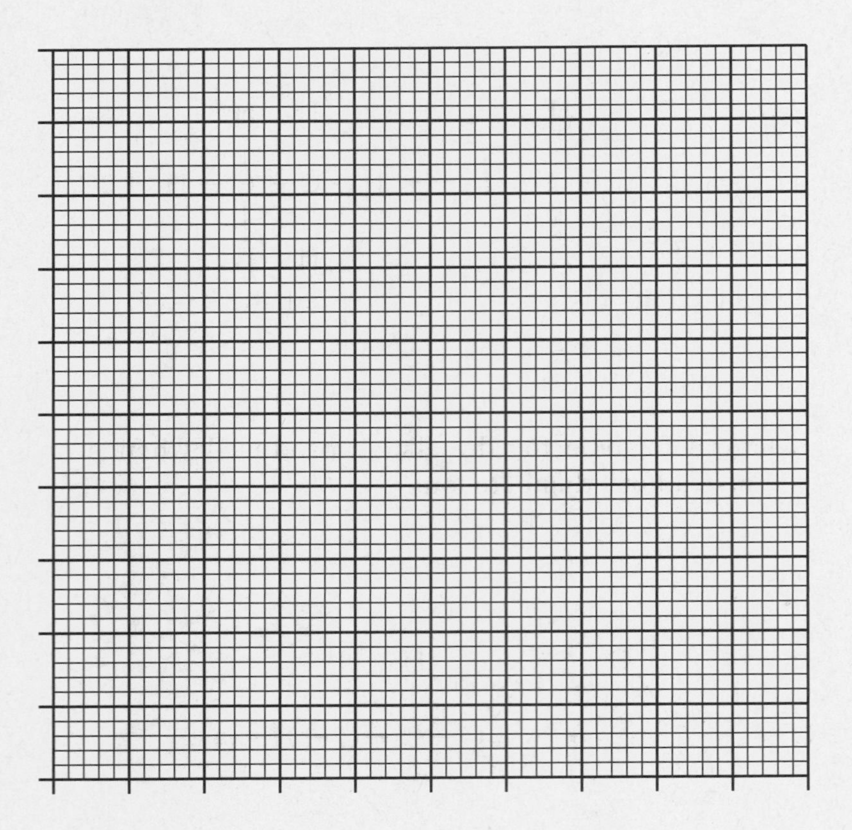

DO NOT WRITE IN THIS MARGIN

| KU | PS |
| --- | --- |

## ADDITIONAL SPACE FOR ANSWERS

DO NOT WRITE IN THIS MARGIN

| KU | PS |
|---|---|

## ADDITIONAL SPACE FOR ANSWERS

STANDARD GRADE | CREDIT

2010

[BLANK PAGE]

FOR OFFICIAL USE

| | | | | | |
|---|---|---|---|---|---|
| | | | | | |

**C**

| | KU | PS |
|---|---|---|
| Total Marks | | |

# 0500/402

NATIONAL QUALIFICATIONS 2010

FRIDAY, 30 APRIL
10.50 AM – 12.20 PM

# CHEMISTRY
## STANDARD GRADE
### Credit Level

**Fill in these boxes and read what is printed below.**

Full name of centre

Town

Forename

Surname

Date of birth
Day Month Year

Scottish candidate number

Number of seat

1 All questions should be attempted.

2 Necessary data will be found in the Data Booklet provided for Chemistry at Standard Grade and Intermediate 2.

3 The questions may be answered in any order but all answers are to be written in this answer book, and must be written clearly and legibly in ink.

4 Rough work, if any should be necessary, as well as the fair copy, is to be written in this book.
Rough work should be scored through when the fair copy has been written.

5 Additional space for answers and rough work will be found at the end of the book.

6 The size of the space provided for an answer should not be taken as an indication of how much to write. It is not necessary to use all the space.

7 Before leaving the examination room you must give this book to the Invigilator. If you do not, you may lose all the marks for this paper.

SA 0500/402 6/23910

## PART 1

**In Questions 1 to 9 of this part of the paper, an answer is given by circling the appropriate letter (or letters) in the answer grid provided.**

**In some questions, two letters are required for full marks.**

**If more than the correct number of answers is given, marks will be deducted.**

**A total of 20 marks is available in this part of the paper.**

### SAMPLE QUESTION

| A | B | C |
|---|---|---|
| $CH_4$ | $H_2$ | $CO_2$ |
| **D** | **E** | **F** |
| $CO$ | $C_2H_5OH$ | $C$ |

(*a*) Identify the hydrocarbon.

| (A) | B | C |
|---|---|---|
| D | E | F |

The one correct answer to part (*a*) is A. This should be circled.

(*b*) Identify the **two** elements.

| A | (B) | C |
|---|---|---|
| D | E | (F) |

As indicated in this question, there are **two** correct answers to part (*b*). These are B and F. Both answers are circled.

If, after you have recorded your answer, you decide that you have made an error and wish to make a change, you should cancel the original answer and circle the answer you now consider to be correct. Thus, in part (*a*), if you want to change an answer A to an answer D, your answer sheet would look like this:

| ~~(A)~~ | B | C |
|---|---|---|
| (D) | E | F |

If you want to change back to an answer which has already been scored out, you should enter a tick (✓) in the box of the answer of your choice, thus:

| ✓ ~~(A)~~ | B | C |
|---|---|---|
| ~~(D)~~ | E | F |

DO NOT WRITE IN THIS MARGIN

Marks | KU | PS

**1.** Crude oil can be separated into fractions.

crude oil →

| Fraction | Number of carbon atoms per molecule |
|---|---|
| A | 1–4 |
| B | 4–10 |
| C | 10–16 |
| D | 16–20 |
| E | 20+ |

(*a*) Identify the fraction which is the most viscous.

| A |
|---|
| B |
| C |
| D |
| E |

**1**

(*b*) Identify the fraction used as camping gas.

| A |
|---|
| B |
| C |
| D |
| E |

**1**

**(2)**

**[Turn over**

DO NOT WRITE IN THIS MARGIN

*Marks* KU PS

**2.** The grid contains the symbols for some common elements.

| A | B | C |
|---|---|---|
| H | N | Si |
| D | E | F |
| Al | Mg | O |

(*a*) Identify the element which has a density of 1·74 g/cm$^3$.

You may wish to use the data booklet to help you.

| A | B | C |
|---|---|---|
| D | E | F |

**1**

(*b*) Identify the **two** elements which react together to form a molecule with the same shape as a methane molecule.

| A | B | C |
|---|---|---|
| D | E | F |

**1**

(*c*) Identify the **two** elements which form an ionic compound with a formula of type $\mathbf{X_2Y_3}$, where **X** is a metal.

| A | B | C |
|---|---|---|
| D | E | F |

**1**

**(3)**

DO NOT WRITE IN THIS MARGIN

Marks | KU | PS

3. The grid shows information about some particles.

| A | B | C |
|---|---|---|
| $^{23}_{11}Na$ | $^{18}_{8}O$ | $^{40}_{19}K^{+}$ |
| D | E | F |
| $^{24}_{12}Mg^{2+}$ | $^{35}_{17}Cl^{-}$ | $^{16}_{8}O$ |

(*a*) Identify the **two** particles with the same number of neutrons.

| A | B | C |
|---|---|---|
| D | E | F |

1

(*b*) Identify the particle which has the same electron arrangement as neon.

| A | B | C |
|---|---|---|
| D | E | F |

1

**(2)**

**[Turn over**

DO NOT WRITE IN THIS MARGIN

*Marks* KU PS

**4.** The grid shows the names of some carbohydrates.

| A | fructose |
|---|---|
| B | glucose |
| C | maltose |
| D | sucrose |
| E | starch |

(*a*) Identify the condensation polymer.

| A |
|---|
| B |
| C |
| D |
| E |

**1**

(*b*) Identify the **two** monosaccharides.

| A |
|---|
| B |
| C |
| D |
| E |

**1**

**(2)**

*Marks* KU PS

**5.** Iron can be coated with different materials which provide a physical barrier against corrosion.

| A | oil |
|---|---|
| B | zinc |
| C | plastic |
| D | tin |
| E | paint |

(*a*) Identify the coating which is used to galvanise iron.

| A |
|---|
| B |
| C |
| D |
| E |

**1**

(*b*) Identify the coating which, if scratched, would cause the iron to rust faster than normal.

| A |
|---|
| B |
| C |
| D |
| E |

**1**

**(2)**

**[Turn over**

DO NOT WRITE IN THIS MARGIN

*Marks* | KU | PS

**6.** The structures of some hydrocarbons are shown in the grid.

| A | B | C |
|---|---|---|
| H    H H<br>C=C–C–H<br>H      H | H   H<br>C<br>H–C——C–H<br>H    H | H  H  H<br>H–C–C–C–H<br>H  H  H |
| **D** | **E** | **F** |
| H  H  H  H<br>H–C–C–C–C–H<br>H  H  H  H | H  H  H    H<br>H–C–C–C=C<br>H  H       H | H  H<br>H–C–C–H<br>H–C–C–H<br>H  H |

(*a*) Identify the **two** hydrocarbons with the general formula $C_nH_{2n}$, which do **not** react quickly with bromine solution.

| A | B | C |
|---|---|---|
| D | E | F |

**1**

(*b*) Identify the hydrocarbon which is the first member of a homologous series.

| A | B | C |
|---|---|---|
| D | E | F |

**1**

(*c*) Identify the **two** isomers of

H  H     H<br>
H–C–C=C–C–H<br>
H      H  H

| A | B | C |
|---|---|---|
| D | E | F |

**1**

**(3)**

DO NOT WRITE IN THIS MARGIN

Marks | KU | PS

**7.** Elements can be used in different ways.

| A | B | C |
|---|---|---|
| chlorine | potassium | platinum |
| **D** | **E** | **F** |
| hydrogen | neon | iron |

(*a*) Identify the element which is a reactant in the Haber Process.

| A | B | C |
|---|---|---|
| D | E | F |

1

(*b*) Identify the element used as the catalyst in the manufacture of nitric acid (Ostwald Process).

| A | B | C |
|---|---|---|
| D | E | F |

1

**(2)**

**[Turn over**

DO NOT WRITE IN THIS MARGIN

*Marks* KU PS

**8.** The grid shows some statements which can be applied to different solutions.

| | |
|---|---|
| A | It has a pH less than 7. |
| B | It conducts electricity. |
| C | It contains less $OH^{-}(aq)$ ions than pure water. |
| D | It does not neutralise dilute hydrochloric acid. |
| E | When diluted the concentration of $OH^{-}(aq)$ ions decreases. |

Identify the **two** statements which are correct for an alkaline solution.

| |
|---|
| A |
| B |
| C |
| D |
| E |

**(2)**

DO NOT WRITE IN THIS MARGIN

*Marks* KU PS

**9.** The grid shows pairs of chemicals.

| A copper carbonate + dilute sulphuric acid | B lead nitrate solution + potassium iodide solution | C potassium hydroxide + nitric acid |
|---|---|---|
| D copper + water | E silver + hydrochloric acid | F ammonium nitrate + sodium hydroxide |

Which **two** boxes contain a pair of chemicals that react together to form a gas?

| A | B | C |
|---|---|---|
| D | E | F |

**(2)**

**[Turn over**

Marks | KU | PS

## PART 2

**A total of 40 marks is available in this part of the paper.**

**10.** Poly(methyl methacrylate) is a synthetic polymer used to manufacture perspex.

(*a*) What is meant by the term **synthetic**?

________________________________________ **1**

(*b*) The structure of the methyl methacrylate monomer is shown.

```
H     CH3
|     |
C  =  C
|     |
H     COOCH3
```

methyl methacrylate

(i) Draw a section of the poly(methyl methacrylate) polymer, showing three monomer units joined together.

**1**

(ii) Name the type of polymerisation taking place.

________________________________________ **1**

(*c*) Name a toxic gas produced when poly(methyl methacrylate) burns.

____________________ **1**

**(4)**

DO NOT WRITE IN THIS MARGIN

*Marks* KU PS

**11.** A student set up an experiment to investigate the breakdown of glucose to form alcohol.

At the start a deflated balloon was attached to the top of the tube.

After two hours the balloon inflates as shown.

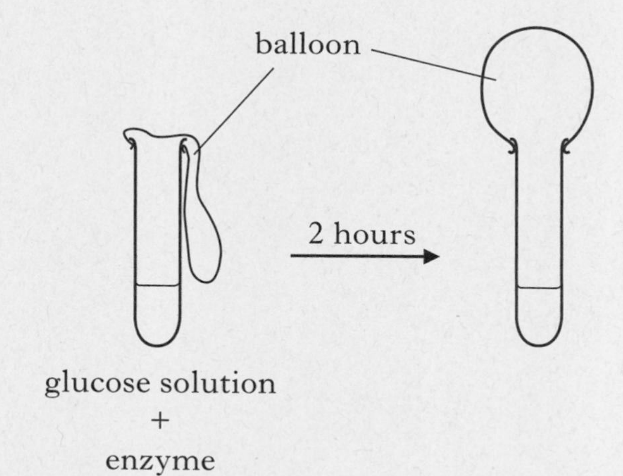


(*a*) (i) Name the type of chemical reaction taking place in the test tube.

______________________________ **1**

(ii) Name the gas produced, which causes the balloon to inflate.

______________________________ **1**

(*b*) The student repeated the experiment at 80 °C.

What effect would this have on how much the balloon inflates?

______________________________

______________________________

______________________________ **1**

**(3)**

**[Turn over**

DO NOT WRITE IN THIS MARGIN

*Marks* KU PS

**12.** A student added magnesium ribbon to an excess of dilute sulphuric acid and measured the volume of hydrogen gas produced.

The reaction stopped when all the magnesium was used up.

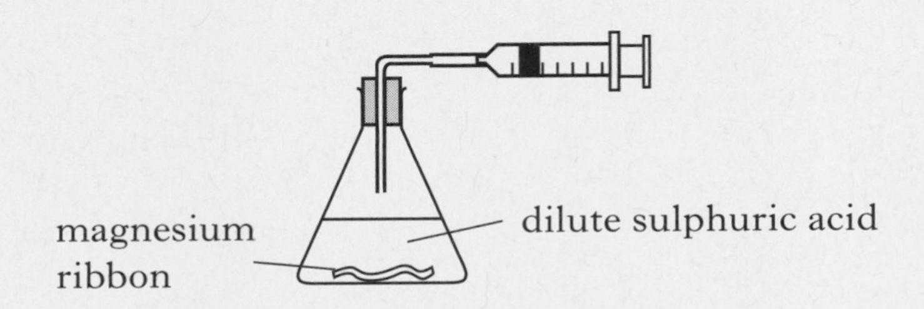


The results are shown in the table.

| **Time/s** | 0 | 10 | 20 | 40 | 50 | 60 | 70 |
|---|---|---|---|---|---|---|---|
| **Volume of hydrogen gas/cm$^3$** | 0 | 20 | 32 | 50 | 52 | 53 | 53 |

(*a*) State the test for hydrogen gas.

________________________________________

________________________________________ **1**

(*b*) Draw a line graph of the results.

*Use appropriate scales to fill most of the graph paper.*

(Additional graph paper, if required, will be found on page 24.)

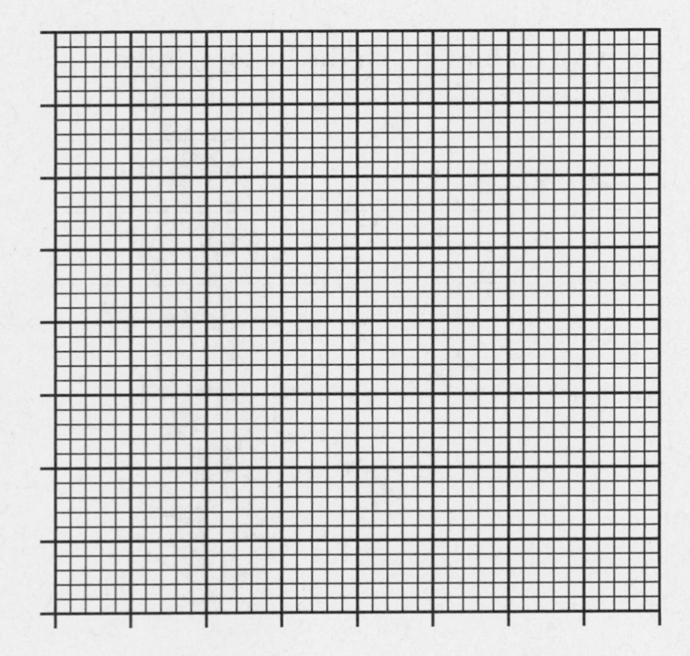

**2**

DO NOT WRITE IN THIS MARGIN

*Marks* KU PS

**12. (continued)**

(*c*) Using your graph, predict the volume of hydrogen gas produced during the first 30 seconds.

____________ $cm^3$ **1**

(*d*) The student repeated the experiment using a higher concentration of acid. The same volume of acid and the same mass of magnesium ribbon were used.

What volume of hydrogen gas would have been produced after 60 seconds?

____________ $cm^3$ **1**

(*e*) Calculate the mass of hydrogen produced when 4·9 g of magnesium reacts with an excess of dilute sulphuric acid.

$$Mg + H_2SO_4 \longrightarrow MgSO_4 + H_2$$

____________ g **2**

**(7)**

**[Turn over**

DO NOT WRITE IN THIS MARGIN

*Marks* KU PS

**13.** A student set up the following experiment to investigate the colour of ions in nickel(II) chromate solution.

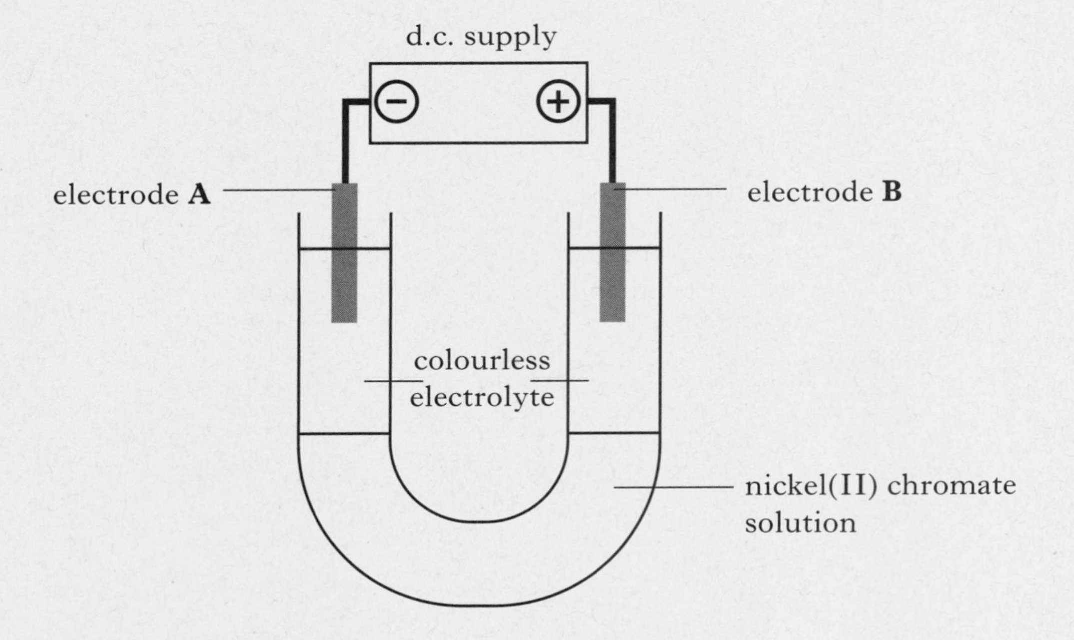


The results are shown.

Green colour moves towards electrode **A**
Yellow colour moves towards electrode **B**

(*a*) Why **must** a d.c. supply be used?

________________________________________

________________________________________

________________________________________ **1**

(*b*) State the colour of the nickel(II) ions.

____________________ **1**

(*c*) Write the **ionic** formula for nickel(II) chromate.

**1**

**(3)**

DO NOT WRITE IN THIS MARGIN

*Marks* KU PS

**14.** The Eurofighter "Typhoon" is made from many newly developed materials including titanium alloys.

(*a*) The first step in extracting titanium from its ore is to convert it into titanium(IV) chloride.

Titanium(IV) chloride is a liquid at room temperature and does **not** conduct electricity.

What type of bonding, does this suggest, is present in titanium(IV) chloride?

____________________________ **1**

(*b*) Titanium(IV) chloride is then reduced to titanium metal.

The equation for the reaction taking place is:

$$TiCl_4 \quad + \quad Na \longrightarrow Ti \quad + \quad NaCl$$

(i) Balance the equation. **1**

(ii) What does this reaction suggest about the reactivity of titanium compared to that of sodium?

________________________________________________________ **1**

**(3)**

**[Turn over**

DO NOT WRITE IN THIS MARGIN

*Marks* KU PS

**15.** Scuba divers can suffer from painful and potentially fatal problems if they rise to the surface of the water too quickly. This causes dissolved nitrogen in their blood to form bubbles of nitrogen gas.

| Distance from surface of water/m | Concentration of dissolved nitrogen/units |
|---|---|
| 0 | 11·5 |
| 10 | 23·0 |
| 20 | 34·5 |
| 30 | 46·0 |
| 40 | 57·5 |

(*a*) Describe the relationship between the distance from the surface of the water and the concentration of dissolved nitrogen.

______________________________________________

______________________________________________

______________________________________________ **1**

(*b*) Predict the concentration of dissolved nitrogen at 60 m.

______________ units **1**

(*c*) A nitrogen molecule is held together by three covalent bonds.

Circle the correct words to complete the sentence.

In a covalent bond the atoms are held together by the attraction between the positive {electrons / neutrons / protons} and the shared pair of negative {electrons / neutrons / protons}. **1**

**(3)**

DO NOT WRITE IN THIS MARGIN

Marks KU PS

**16.** (*a*) Galena is an ore containing lead sulphide, PbS.

(i) What is the charge on this lead ion?

______________________ **1**

(ii) Calculate the percentage by mass of lead in galena, PbS.

__________ % **2**

(*b*) Most metals have to be extracted from their ores.

(i) Name the metal extracted in a Blast furnace.

______________________ **1**

(ii) Place the following metals in the correct space in the table.

copper, mercury, aluminium

You may wish to use the data booklet to help you.

| **Metal** | **Method of extraction** |
|---|---|
| | using heat alone |
| | electrolysis of molten ore |
| | heating with carbon |

**1**

**(5)**

**[Turn over**

DO NOT WRITE IN THIS MARGIN

*Marks* KU PS

**17.** Iron displaces silver from silver(I) nitrate solution.

The equation for the reaction is:

$$Fe(s) + 2Ag^{+}(aq) + 2NO_3^{-}(aq) \longrightarrow Fe^{2+}(aq) + 2Ag(s) + 2NO_3^{-}(aq)$$

(*a*) Circle the spectator ion in the above equation. **1**

(*b*) Describe a chemical test, including the result, to show that $Fe^{2+}(aq)$ ions are formed.

______________________________________________

______________________________________________ **1**

(*c*) Write the ion-electron equation for the **reduction** step in the reaction.

You may wish to use the data book to help you.

**1**

(*d*) This reaction can also be carried out in a cell.

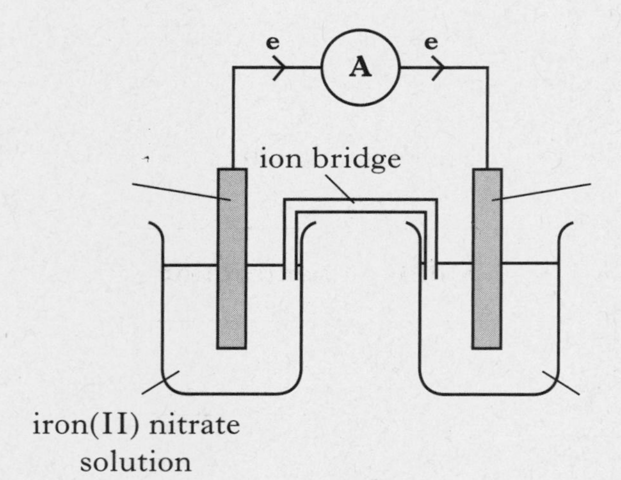


Complete the three labels on the diagram. **1**

(An additional diagram, if required, will be found on page 24.) **(4)**

DO NOT WRITE IN THIS MARGIN

*Marks* KU PS

**18.** Ethers are useful chemicals.

Some are listed in the table.

| Structural formula | Name of ether |
|---|---|
| $CH_3CH_2 - O - CH_2CH_3$ | ethoxyethane |
| $CH_3 - O - CH_2CH_2CH_3$ | methoxypropane |
| $CH_3 - O - CH_2CH_3$ | methoxyethane |
| $CH_3CH_2 - O - CH_2CH_2CH_3$ | **X** |

(*a*) Suggest a name for ether **X**.

______________________________ **1**

(*b*) The boiling points of ethers and alkanes are approximately the same when they have a **similar** relative formula mass.

Suggest the **boiling point** of ethoxyethane (relative formula mass 74).

You may wish to use the data booklet to help you.

__________ °C **1**

**(2)**

**[Turn over**

DO NOT WRITE IN THIS MARGIN

*Marks* | KU | PS

**19.** A student carried out a titration using the chemicals and apparatus below.

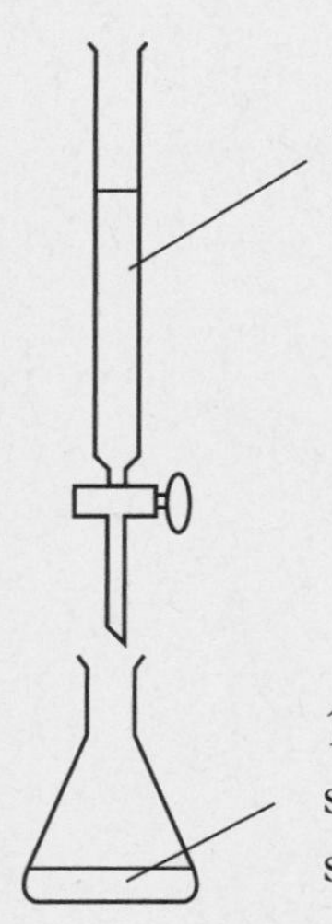


| | Rough titre | 1st titre | 2nd titre |
|---|---|---|---|
| **Initial burette reading/$cm^3$** | 0·3 | 0·2 | 0·5 |
| **Final burette reading/$cm^3$** | 26·6 | 25·3 | 25·4 |
| **Volume used/$cm^3$** | 26·3 | 25·1 | 24·9 |

(*a*) Using the results in the table, calculate the **average** volume of hydrochloric acid required to neutralise the sodium hydroxide solution.

____________ $cm^3$ **1**

(*b*) The equation for the reaction is:

$$HCl + NaOH \longrightarrow H_2O + NaCl$$

Using your answer from part (*a*), calculate the concentration of the sodium hydroxide solution.

**Show your working clearly.**

____________ mol/l **2**

**(3)**

DO NOT WRITE IN THIS MARGIN

Marks | KU | PS

**20.** Chemists have discovered a way to insert a $-CH_2-$ group into any bond which includes an atom of hydrogen.

When a $-CH_2-$ group is inserted into a methanol molecule the following reaction takes place.

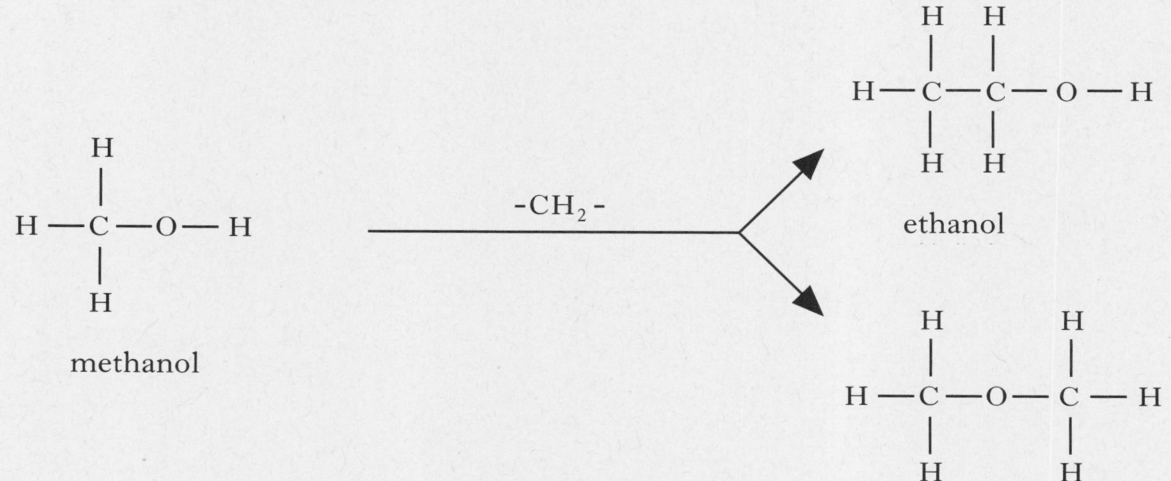


(*a*) This reaction can be repeated using ethanol.

One of the products for this reaction is shown.

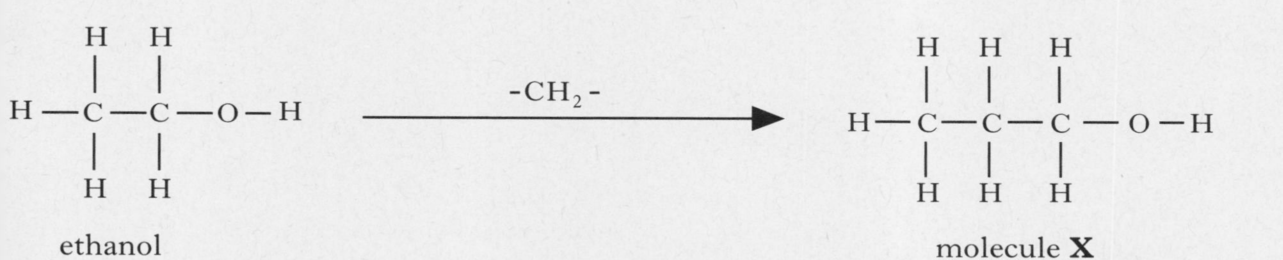


(i) Suggest a name for molecule **X**.

________________________ **1**

(ii) Draw a structural formula for another molecule which would be formed in this reaction.

**1**

(*b*) Identify the **two** products formed when molecule **X** is completely burned in a plentiful supply of oxygen.

________________________________________________ **1**

**(3)**

[*END OF QUESTION PAPER*]

DO NOT WRITE IN THIS MARGIN

| KU | PS |
|---|---|

## ADDITIONAL SPACE FOR ANSWERS

ADDITIONAL GRAPH PAPER FOR QUESTION 12(*b*)

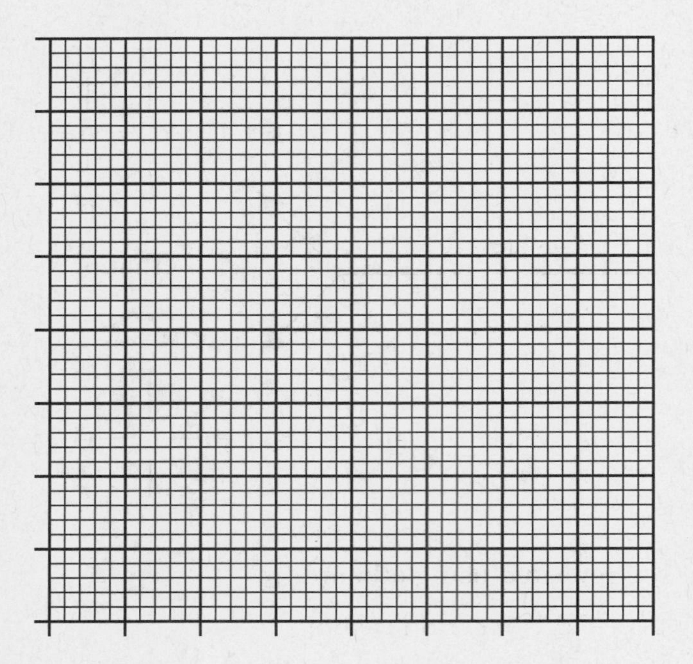

ADDITIONAL DIAGRAM FOR QUESTION 17(*d*)

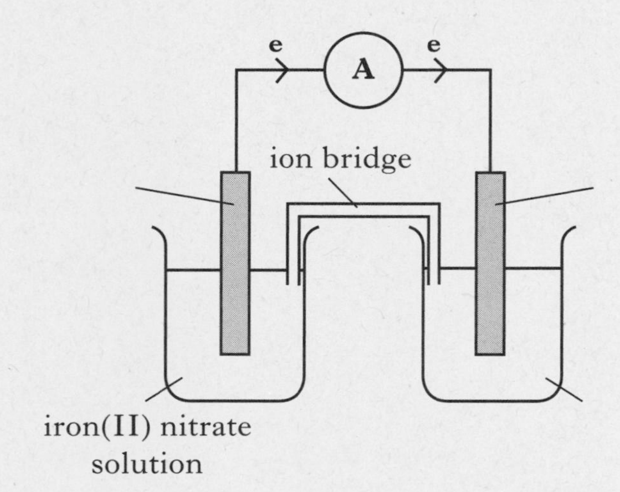

DO NOT WRITE IN THIS MARGIN

| KU | PS |
|---|---|

## ADDITIONAL SPACE FOR ANSWERS

DO NOT WRITE IN THIS MARGIN

| KU | PS |
|---|---|

## ADDITIONAL SPACE FOR ANSWERS

[BLANK PAGE]

[BLANK PAGE]